T0327379

Welding Engineering

Welding Engineering

An Introduction

David H. Phillips
The Ohio State University
OH, USA

Second Edition

Trademarks
Wiley and the Wiley logo are trademarks or registered trademarks of John Wiley & Sons, Inc. and/or its affiliates in the United States and other countries and may not be used without written permission. All other trademarks are the property of their respective owners. John Wiley & Sons, Inc. is not associated with any product or vendor mentioned in this book.

Limit of Liability/Disclaimer of Warranty
While the publisher and author have used their best efforts in preparing this book, they make no representations or warranties with respect to the accuracy or completeness of the contents of this book and specifically disclaim any implied warranties of merchantability or fitness for a particular purpose. No warranty may be created or extended by sales representatives or written sales materials. The advice and strategies contained herein may not be suitable for your situation. You should consult with a professional where appropriate. Neither the publisher nor author shall be liable for any loss of profit or any other commercial damages, including but not limited to special, incidental, consequential, or other damages. Further, readers should be aware that websites listed in this work may have changed or disappeared between when this work was written and when it is read. Neither the publisher nor authors shall be liable for any loss of profit or any other commercial damages, including but not limited to special, incidental, consequential, or other damages.

For general information on our other products and services or for technical support, please contact our Customer Care Department within the United States at (800) 762-2974, outside the United States at (317) 572-3993 or fax (317) 572-4002.

Wiley also publishes its books in a variety of electronic formats. Some content that appears in print may not be available in electronic formats. For more information about Wiley products, visit our web site at www.wiley.com.

A catalogue record for this book is available from the Library of Congress
Hardback ISBN: 9781119858720; epub ISBN: 9781119858744; ePDF ISBN: 9781119858737

Cover Images: © wi6995/Adobe Stock Photos; Jenson/Shutterstock
Cover design by Wiley

Set in 9.5/12.5pt STIXTwoText by Integra Software Services Pvt. Ltd, Pondicherry, India

"To my dad, David M. Phillips, a perpetual source of encouragement who loved everything about The Ohio State University, and would have been so proud to know that his son became a professor there."

Contents

Preface

The Welding Engineering program at The Ohio State University is well known for its long history of graduating students who serve critical roles in industry. This textbook was written for use in an undergraduate course in Welding Engineering that I have taught at Ohio State for the past 12 years. The course serves as an introduction to the Welding Engineering curriculum at Ohio State, and is intended to prepare sophomore students for more in-depth Welding Engineering courses in welding processes, metallurgy, design, and NDE, which are required courses they take during their junior and senior years. Much of what is included in this book comes from my class lectures. Since both the course and this book represent "An Introduction" to all the important topics associated with the field of Welding Engineering, the coverage of each of the topics is intended to be relatively brief and concise. Fundamentals and basic concepts are emphasized, while many of the details are intentionally omitted. So, while it is not intended to serve as a handbook, recommended reading for further information and greater detail is provided at the end of each chapter.

This second edition includes many new sections and updates. All chapters (except the first introductory chapter) now end with a self-assessment titled "Test Your Knowledge." The self-assessment consists of several simple True/False questions covering important topics in the chapter, as well as a more involved "solve a welding engineering problem" question. This question asks the student to apply knowledge gained from that chapter toward solving a typical welding engineering challenge. In addition to the numerous chapter updates and revisions, many new topics have been added, including additive manufacturing, computational modeling of welds, advanced high strength steels, and precipitation-hardening stainless steels. Although this book is intended for sophomore-level Welding Engineering students, it should also serve as a useful guide to other engineers, technicians, and specialists who are working in the field of welding and are seeking a more fundamental understanding of the important concepts. Universities offering Welding Engineering Technology programs may want to consider it as well.

About the Companion Website

This book is accompanied by a companion website:

www.wiley.com/go/Phillips/WeldingEngineeringIntroduction

This website includes

- Answer Key

1

What Is Welding Engineering?

Welding Engineering is a complex field that requires proficiency in a broad range of engineering disciplines. Students who pursue a degree in Welding Engineering engage in a curriculum that is more diverse than other engineering disciplines (Figure 1.1). They take advanced courses in welding metallurgy and materials science that cover materials ranging from carbon steels and stainless steels, to nonferrous alloys such as nickel, aluminum, and titanium, as well as polymers. Welding process courses emphasize theory, principles, and fundamental concepts pertaining to the multitude of important industrial welding processes.

While many associate welding with arc welding processes, a Welding Engineer may be responsible for many other processes throughout their career. Therefore, in addition to arc welding, the Welding Engineering curriculum includes thorough coverage of high-energy density processes such as Laser and Electron Beam Welding, solid-state welding processes such as Friction Welding and Explosion Welding, and resistance welding processes including Spot and Projection Welding. Students are trained in many important electrical concepts associated with welding such as process control and transformer theory and operation. Welding design courses cover the principles of important subjects such as heat flow, residual stress, fatigue and fracture, weld sizing, and weld design for various loading conditions. Analysis through computational modeling is included in many of the courses. Nondestructive testing techniques including x-ray, ultrasonics, eddy current, magnetic particle, and dye penetrant are emphasized as well.

The diverse Welding Engineering curriculum prepares its graduates for a wide range of possible career paths and industrial fields. Working environments include automation and high-speed production, fabrication, manufacturing, and research. Welding Engineering graduates are typically in high demand and choose jobs from a variety of industry sectors including nuclear, petrochemical, automotive, medical, shipbuilding, aerospace, power generation, and heavy equipment sectors.

1.1 Introduction to Welding Processes

Considering the recent developments in hybrid approaches to welding, there are probably now more than 75 types of welding processes available for the manufacturer or fabricator to choose from. The reason that there are so many processes is that each process has its own unique advantages and disadvantages that make them ideal for some applications and a poor choice for others. Arc welding processes offer advantages such as portability and low cost but are relatively slow and rely on a considerable amount of heating to produce the weld. High-energy density processes such

Welding Engineering: An Introduction, Second Edition. David H. Phillips.
© 2023 John Wiley & Sons, Inc. Published 2023 by John Wiley & Sons, Inc.
Companion Website: www.wiley.com/go/Phillips/WeldingEngineeringIntroduction

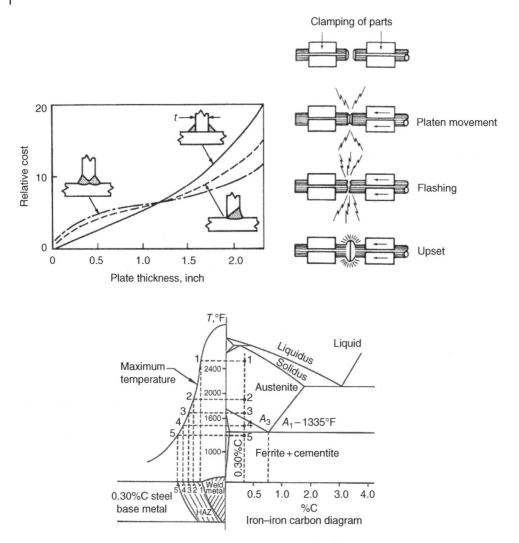

Figure 1.1 A sampling of welding engineering topics—design (top left), processes (top right), and welding metallurgy (bottom). Adapted from AWS Handbook.

as Laser Welding produce low heat inputs and fast welding speeds, but the equipment is very expensive and joint fit-up needs to be nearly perfect. Solid-state welding processes avoid many of the weld discontinuities associated with melting and solidification, but are also very expensive and often are restricted to limited joint designs. Resistance welding processes are typically very fast and require no additional filler materials, but are often limited to thin sheet applications or very high production applications such as the seams in welded pipe.

Each of these processes produces a weld (metallic bond) using some combination of heat, time, and/or pressure. Those that rely on extreme heat at the source such as arc and high-energy density processes generally need no pressure. A process such as Diffusion Welding relies on some heating and some pressure, but with a considerable amount of time. Explosion Welding relies on a tremendous amount of pressure, with minimal heating and time to produce the weld.

When choosing an optimum process for a given application, the Welding Engineer must consider all the above, including much more that will be covered in the next few chapters on welding processes.

2

Arc Welding Processes

2.1   Fundamentals and Principles of Arc Welding

This section serves as a general introduction to all the arc welding processes. The common features, important concepts, and terminology of this family of welding processes are reviewed, with more process-specific details provided in the sections that follow. Arc welding refers to a family of processes that rely on the extreme heat of an electric arc to create a weld. They usually, but not always, involve the use of additional filler metal to produce the weld. As one of the first welding processes, arc welding continues to be very popular primarily due to its low equipment cost, portability, and flexibility. Some of the key developments that led to modern arc welding include the discovery of the electric arc in the 1820s (Davies), the first welding patent using a carbon electrode in 1886, the first covered electrode in 1900 (Kjellberg), and the first process using a continuously fed electrode in the 1940s.

The most common arc welding processes today are charted in Figure 2.1. The abbreviations refer to the American Welding Society (AWS) terminology as follows: SMAW—Shielded Metal Arc Welding, GMAW—Gas Metal Arc Welding, GTAW—Gas Tungsten Arc Welding, PAW—Plasma Arc Welding, SAW—Submerged Arc Welding, FCAW—Flux Cored Arc Welding, SW—Arc Stud Welding, and EGW—Electrogas Welding. Although technically not an arc welding process, ESW—Electroslag Welding is very similar to EGW and as such, is often included with arc welding. In practice, older designations and trade names of processes are often used, some of which are given in italics in the figure. Examples are Stick or Covered Electrode welding for SMAW, MIG meaning Metal Inert Gas for GMAW, and TIG meaning Tungsten Inert Gas for GTAW. One key generalization in the modern terms is the substitution of "G" for "IG" denoting inert gas since these processes no longer rely solely on inert gasses for shielding.

With all arc welding processes, the initiation of an arc basically completes (or closes) an electrical circuit. As shown in Figure 2.2, the most basic arc welding arrangement consists of an arc welding power supply, electrode, work cables (or leads), means to connect to the electrode (electrode holder with SMAW as shown), and the work piece or parts to be welded. A range of typical currents and voltages are shown. Open circuit voltages provided by traditional power supplies are in the range of 60–80 volts although they are sometimes lower. The open circuit voltage is the voltage between the electrode and ground clamp after the power supply is turned on, but before the arc is initiated. This voltage range is high enough to establish and maintain an arc, but low enough to minimize the risk of electrical shock. Once the arc is established, the voltage across the arc typically ranges between 10 volts and 40 volts, depending on the process.

Welding Engineering: An Introduction, Second Edition. David H. Phillips.
© 2023 John Wiley & Sons, Inc. Published 2023 by John Wiley & Sons, Inc.
Companion Website: www.wiley.com/go/Phillips/WeldingEngineeringIntroduction

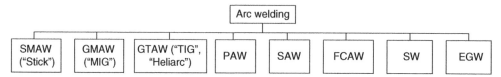

Figure 2.1 Common arc welding processes.

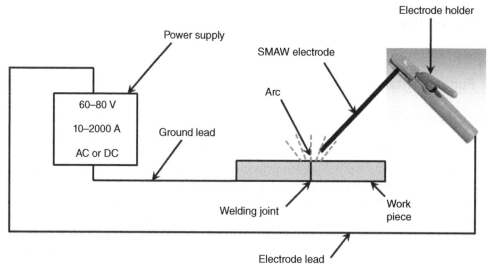

Figure 2.2 Arc welding circuit depicting SMAW.

Welding power supplies are usually designed to deliver direct current electricity referred to as DC. A pulsing output called pulsed direct current has become a prominent feature in many advanced welding power supplies. Programmable pulsing parameters or preprogrammed pulsing schedules can be used to optimize welding performance, primarily for GMAW. Alternating current (AC) is sometimes used. One benefit is that AC machines are simple and inexpensive. Welding with AC is also a very effective way to weld aluminum, which will be discussed later in the section on GTAW. A form of pulsing known as variable polarity is another advanced capability of many modern power supplies. Variable polarity capability allows for the customization of pulsing frequency and waveform to optimize welding performance.

Arc welding processes can be divided into three categories: manual, semiautomatic, and automatic. Manual processes such as SMAW and GTAW require the welder to control travel speed, arc length, and electrode (or filler metal) feed rate. As a result, these processes require the most welder skill. Semiautomatic processes such as GMAW use a continuous motorized wire feed mechanism. The power supply controls the arc length through a concept known as "self-regulation" to be discussed later. Since the welder only needs to control the position and travel speed of the gun, GMAW is relatively easy to learn, and for the same reasons is also an ideal process for robotic welding. SAW is an example of an automatic arc welding process because all the variables of travel speed, arc length, and electrode feed are controlled by the machine. Therefore, there is no welding skill required to operate automatic processes, only the knowledge needed to set up the machine and select the proper variables. Any arc welding process may be referred to as mechanized or automated if it is attached to a travel mechanism or robotic arm.

2.1.1 Fundamentals of an Electric Arc

An electric arc is a type of electrical discharge that occurs between electrodes when a sufficient voltage is applied across a gap causing the gas to break down or ionize (Figure 2.3). Gas is normally an insulator, but once ionized it becomes a conductor of electricity. Ionization occurs when the gas atoms lose bound electrons that are then free to travel independently in the gas to produce an electric current. These free electrons pick up energy from the electric field produced by the applied voltage and collide with other gas atoms. This allows the ionization process to grow resulting in an "avalanche" effect. Once the gas is highly ionized, it becomes relatively easy for electrons to flow, and under the right conditions a stable electric arc can be formed.

An ionized gas consists of free electrons that flow in one direction and positive ions that flow in the other direction. Collisions with mostly neutral atoms produce a tremendous resistive heating of the gas, so, in a sense, the arc is a large resistor. The extreme heat also maintains the ionization process. Electromagnetic radiation is given off due to the high temperatures resulting in the characteristic glow of the arc. In addition to the observable visible wavelengths, large amounts of invisible infrared and ultraviolet wavelengths are emitted. The ionized glowing gas that makes up the arc is often referred to as plasma. For the arc to be maintained, the power supply must be able to supply the high current and low voltage demanded by the arc.

The utility of the electric arc to welding is the extreme heat that is produced under stable arc conditions which can melt most metals and form what is known as a weld pool or puddle. Arc temperatures are known to range from 5000 up to 30,000 C. As Figure 2.4 indicates, the temperature of an arc is hottest at its center since the outer portions of the arc lose heat to the surroundings due to convection, conduction, and radiation. A major contribution of heat to the welding and work electrodes is actually not just the extremely high arc temperatures, but the intense energy dissipative processes at the arc attachment points to the welding and work electrodes. This will be discussed later.

For consumable electrode processes such as SMAW and GMAW, the arc contains molten droplets of filler metal, which melt from the electrode and travel through the arc to the weld pool. As will be discussed later, the size, shape, and way the molten metal travels from the electrode to the puddle are known as the modes of metal transfer. This is of particular interest with the GMAW process but is not an important consideration for the other arc welding processes. Filler metal transfer through the arc inevitably results in some molten drops being ejected from the arc or weld pool that may stick to the

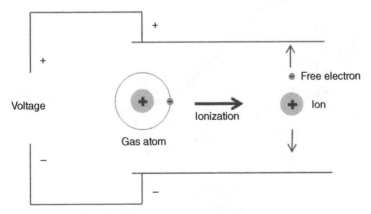

Figure 2.3 Ionization of a gas and current flow in an arc.

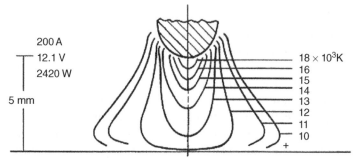

Figure 2.4 Thermal diagram of gas tungsten arc. (*Source:* Reproduced by permission of American Welding Society, ©*Welding Handbook*).

part. This is called spatter and is often a quality concern. GTAW and PAW processes deliver the filler metal directly to the weld puddle (not through the arc) and are therefore not susceptible to spatter.

2.1.2 Arc Voltage

It was mentioned previously that operating arc voltages typically fall in the range of 10–40 volts. Arc voltages are primarily related to arc lengths. Longer arc lengths produce higher arc voltages and shorter arcs produce lower voltages. Figure 2.5 shows how voltage (potential) varies through the arc. As the figure indicates, a significant amount of the voltage distribution or drop across the arc is close to the anode and the cathode. These regions are known as the anode drop or fall at the positive electrode (the work piece in the figure) and the cathode drop or fall at the negative electrode (the welding

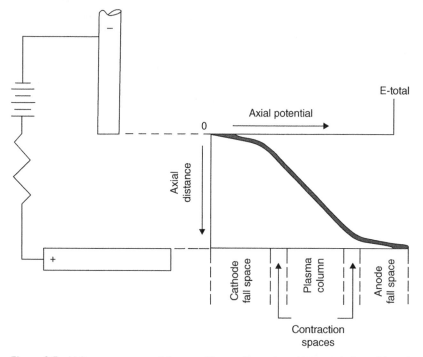

Figure 2.5 Voltage across a welding arc. (*Source:* Reproduced by permission of American Welding Society, ©*Welding Handbook*).

electrode in the figure). The primary change in arc voltage as a function of arc length is known to be associated with the region between the anode and cathode drops called the plasma column. Voltages at the anode and cathode drop regions are not significantly affected by arc length. As a result, even extremely short arc lengths will exhibit voltages much greater than zero. This provides evidence that most of the arc voltage exists at the two voltage drop regions at the electrodes. For a typical welding arc length, these voltages may represent as much as 80–90% of the total arc voltage. Since heat generation and power dissipation are functions of voltage and current, and the level of current is uniform through the arc, the amount of power dissipation must therefore be greatest at the electrode drop regions and not in the plasma column. These anode and cathode drop regions are extremely narrow, and, therefore, their effect is not revealed on thermal diagrams of arcs such as that shown in Figure 2.4. Nevertheless, they play a critical role in the melting at the anode and cathode, which is why the arc temperature alone is not the key to explaining the arc as an effective heat source for welding.

Higher voltages are required to ionize a gas across a given gap and gas pressure. The latter is usually atmospheric pressure for a welding arc; however, arc welding can be conducted at other pressure conditions as well such as under water or in a chamber. Since the open circuit voltage typical of power supplies (60–80 volts) is relatively low, it is not sufficient to simply break down a gap. With manual arc welding processes, it is usually necessary to touch, or so-called scratch or drag the electrode on the work piece. This produces an instantaneous short circuit current from the power supply and is referred to as "striking" or "drawing" an arc. Once the ionization process has been initiated, the gap can be increased to achieve a stable arc. With semiautomatic processes, the wire is driven into the work piece by the wire feed mechanism producing a short circuit. The power supply reacts by producing a very high short circuit current that rapidly melts the wire forming a gap. The differences between arc welding power supplies for manual and semiautomatic processes will be explained later in this chapter. With GTAW, special arc starting systems that produce high voltages at a high frequency may be integrated with the power supply so the welder can initiate the arc without touching the tungsten electrode to the part. Touching the tungsten electrode can produce contamination in the weld or on the electrode tip altering its performance.

2.1.3 Polarity

The electrical polarity applied to the arc via the power supply is very important for the operation of an arc process. The direction of current flow in the arc produces the main effects of polarity on welding. There is potential for confusion regarding the direction of current flow since in welding literature current is commonly described as being in the direction of electron flow, or from the negative to the positive electrode. However, according to standard electrical convention, current is described as flowing from the positive to the negative electrode, or in the direction of the positively charged ions. In any case, in a welding arc, electrons flow from the negative electrode (cathode) to the positive electrode (anode), but this has different effects with different processes. In arc welding, a negative electrode and positive work polarity is referred to as DCEN (DC electrode negative) or historically DCSP (DC straight polarity). The electrons flow out of the welding electrode, through the arc and into the work piece. When the electrode is positive relative to the work the polarity is called DCEP (DC electrode positive) or historically DCRP (DC reverse polarity). Other polarity options are AC (simple alternating current) and VP (variable polarity) which refers to voltage waveforms and frequencies that are more complex and controllable than simple AC. In both cases, the polarity and direction of electron flow alternate during welding.

The effect of polarity on heat input and arc behavior differs with the process and the characteristics of the material being welded. For GTAW, DCEN produces the predominance of heat into the work, and is the most common polarity (Figure 2.6). This is because the tungsten electrode can be heated to extremely high temperatures without melting. At these extremely high temperatures, electrons are easily emitted or "boiled off" from the tungsten electrode (cathode) by a process known as thermionic emission. This produces a stable arc with most of the arc heat concentrating at the work piece where the electrons are deposited. When operating with DCEP polarity, most of the arc heat goes into the electrode which greatly accelerates electrode wear. But DCEP can be beneficial when welding aluminum since the electron emission process can help remove the tenacious aluminum oxide from the surface, a process known as cleaning action. This is where AC current can be advantageous since it delivers a half cycle of DCEN which heats the work piece, and a half cycle of DCEP which removes the oxide but doesn't excessively heat the electrode.

With GMAW, DCEN is generally not usable since the much lower temperature of the melting bare electrode wire cannot easily achieve thermionic emission, resulting in an arc that is very erratic and difficult to control. The unstable and erratic arc is the primary reason DCEN will generate minimal heat into the work piece in the rare cases when it might be used, and in those cases special wire that promotes thermionic emission will be required. On the other hand, DCEP produces a stable arc, and therefore, can sufficiently heat the work piece for welding. The arc stability with DCEP polarity when welding with the GMAW (Figure 2.7) process is due in part to the metal oxides on the work surface which facilitate the electron emission process. In addition to these surface oxides, thermionic emission becomes easier because electrons have a larger surface area when exiting the work piece as compared to the end of an electrode.

Processes where fluxes are used such as SMAW, FCAW, and SAW can use DCEP, DCEN, or AC polarities, depending on the type of flux and the application. Flux additions that are in contact with the welding electrode can promote electron emission when the electrode is the cathode (DCEN). This allows the DCEN polarity to be an effective process choice. In some cases, DCEN may be selected to produce higher deposition rates due to greater electrode heating, with less heat input to

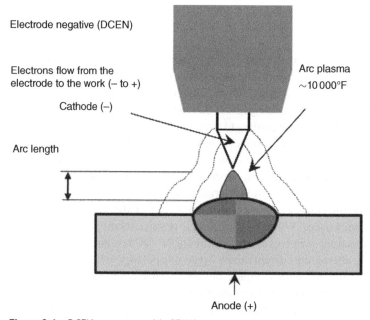

Figure 2.6 DCEN—common with GTAW.

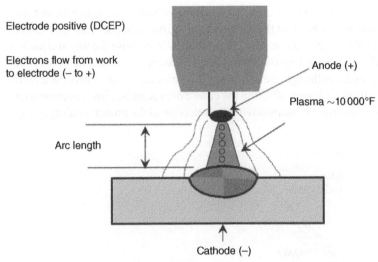

Electrode positive (DCEP)

Electrons flow from work
to electrode (– to +)

Anode (+)

Plasma ~10 000°F

Arc length

Cathode (–)

Figure 2.7 DCEP—common with GMAW.

the part. Because of the reduced heating at the part, with the proper electrode, the DCEN polarity
works very well for welding thinner materials.

2.1.4 Heat Input

The energy or heat input that occurs in the making of an arc weld is an important consideration. It
is expressed as energy per unit length, and is primarily a function of voltage, current, and weld
travel speed as indicated in Figure 2.8. Although voltage plays a prominent role in the equation, it
is a variable that is chosen primarily to create a stable arc and not for affecting heat input. Arc
efficiency, f_1, refers to what percentage of the total heat produced by the arc is delivered to the
weld. Weld heat input is an important consideration because it affects the amount of distortion and
residual stress in the part, and the mechanical properties of the welded part that are a function of
the metallurgical transformations that take place during welding. In general, arc welding processes produce higher heat inputs than many other welding processes because arc heating is not
very concentrated and tends to heat large areas of the work piece.

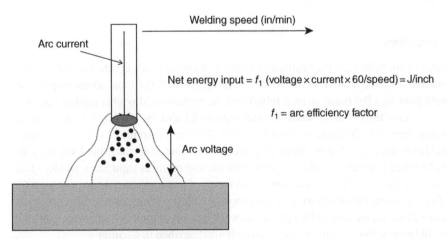

Welding speed (in/min)

Arc current

Net energy input = f_1 (voltage × current × 60/speed) = J/inch

f_1 = arc efficiency factor

Arc voltage

Figure 2.8 Heat input during welding.

Figure 2.9 compares some measured efficiencies (f_1) for different processes. Less efficient processes such as GTAW may lose 50% or more of the arc heat to the surrounding atmosphere. On the other extreme, efficiencies of 90% or greater can be achieved with SAW because the flux and molten slag blanket act as an insulator surrounding the electrode and the arc. Arc efficiencies for other processes lie somewhere between. Although the efficiency of a given arc welding process directly affects heat input, it is typically not a reason affecting the choice of processes. This is because there are usually other much more important reasons driving the selection of the proper welding process for a given application.

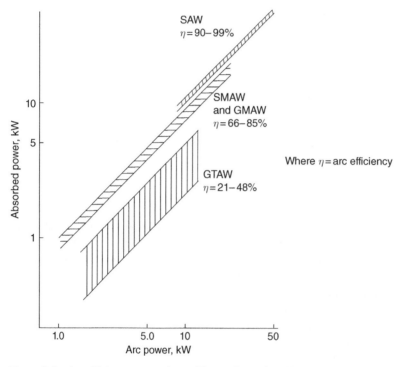

Figure 2.9 Arc efficiency comparisons. (*Source:* Reproduced by permission of American Welding Society, ©*Welding Handbook*).

2.1.5 Welding Position

A significant factor in arc welding is the position of welding. Welding position refers to the way the weld joint is oriented in space relative to the welder. A flat position is the most common position in which the weld joint lays flat (such as on a table) and the molten weld pool is held in the joint by gravity. This is usually the easiest position for making a weld. But the flat position may not be an option, so welds often must be made in other positions. The most extreme is an overhead position that is directly opposite to a flat position. In this case, the molten metal is held solely by its surface tension. Overhead position welding is very difficult and requires significant welder skill. Therefore, welder qualifications include the important factor of position since some positions are much more difficult to master than others. Consideration of welding positions may also affect the choice of processes since not all arc welding processes work in all positions. AWS provides specific designations for all the possible welding positions which are described in Chapter 7.

2.1.6 Filler Metals and Electrodes

All arc welding processes except for GTAW and PAW use a consumable electrode. It is considered consumable because it melts from arc heating and mixes with the molten base metal to create the weld metal. It is called an electrode because it is part of the electrical circuit carrying the current to the arc. The filler metal used for GTAW and PAW is not called an electrode since it does not carry the current. There are numerous varieties of filler metals and electrodes available for different materials, arc welding processes, and applications. AWS provides specifications for filler metals which govern their production, but they are also subject to much proprietary protection regarding exact constituents and formulas.

2.1.7 Shielding

When metals are heated to high temperatures approaching or exceeding their melting point, reactions with the surrounding atmosphere are accelerated and the metals become very susceptible to contamination. Elements that can be most damaging are oxygen, nitrogen, and hydrogen. Contamination from these elements can result in the formation of embrittling phases such as oxides and nitrides, as well as porosity due to entrapment of gasses that form bubbles in the solidifying weld metal. To avoid this contamination, the metal must be shielded as it solidifies and begins to cool. The arc welding processes all rely on either a gas or a flux, or a combination of both for shielding. The way these processes utilize shielding is one of their main distinguishing features. Shielding is important to not only protect the molten metal, but the surrounding heated metal as well. Some metals such as titanium are especially sensitive to contamination from the atmosphere, and often require more thorough shielding techniques.

2.1.7.1 Gas Shielding

Processes such as GMAW, GTAW, and PAW rely solely on externally delivered gas for shielding. Gasses protect by purging atmospheric gasses away from the susceptible metal. The welding gun (or torch for GTAW) is designed so that a coaxial shielding gas flow emanates from the gun, which surrounds the electrode and blankets the weld area. In the case of GTAW and PAW, inert gasses are used with argon being the most common. Helium and blends of helium and argon can also be used. Helium is more expensive than argon, but it transfers more heat from the plasma column to the part due to the higher thermal conductivity of its ionized gas. In some cases, small amounts of hydrogen are added to argon to improve heating by transferring energies of molecular dissociation of the hydrogen in the plasma column to the work.

The GMAW process commonly uses argon for nonferrous metals, particularly aluminum. CO_2 or blends of argon and O_2 or CO_2 are used for ferrous materials such as steels. CO_2 gas produces more spatter and a rougher weld bead appearance, but can produce fast welding speeds, is readily available, and is inexpensive (because it is common and widely used commercially in products such as carbonated beverages). In some cases, additions of CO_2 or small amounts of O_2 to argon can improve electron emission from the negative electrode (or work piece) and enhance weld metal flow by affecting the surface tension of the molten puddle. Helium or blends of argon and helium are sometimes used for nonferrous metals. The choice of shielding gas for GMAW also plays a major role in the mode of molten metal transfer from the electrode to the weld pool.

2.1.7.2 Flux Shielding

The processes of SMAW, SAW, and FCAW use flux for shielding. A welding flux is a material used to prevent or minimize the molten and heated solid metal from forming potentially detrimental constituents such as oxides and nitrides, and to facilitate the removal of such substances if they form or are present prior to welding. Fluxes are used in arc welding processes in three different ways. They are: (1) applied in a granular form to the surface ahead of the weld (SAW), (2) bound with a binding agent to bare electrode wire (SMAW), and (3) contained in the core of a tubular wire (FCAW). Welding fluxes easily absorb moisture and therefore can be a major source of hydrogen to the weld puddle. As a result, when using processes that rely on flux, special care must often be taken when welding certain steels known to be susceptible to hydrogen cracking, a topic that is covered in Chapter 10.

Fluxes provide shielding in two primary ways. With SMAW and FCAW, the fluxes decompose when exposed to the heat of the arc to form CO_2 gas to displace air from the arc. In the case of SAW, the flux melts to form a liquid that reacts with impurities in or on the weld metal to form a slag that floats on the top of the weld pool and later solidifies. SMAW and FCAW also typically form slag in addition to gas, with some electrodes forming more slag than others. Slags are removed at the end of welding and between welding passes with wire brushing or grinding. With some versions of FCAW, shielding gas is delivered through the nozzle as it is with GMAW. Figure 2.10 shows the various ways shielding is implemented with most common arc welding processes.

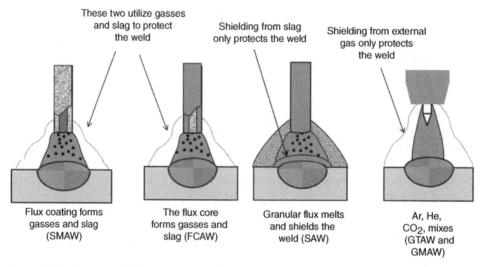

These two utilize gasses and slag to protect the weld

Shielding from slag only protects the weld

Shielding from external gas only protects the weld

| Flux coating forms gasses and slag (SMAW) | The flux core forms gasses and slag (FCAW) | Granular flux melts and shields the weld (SAW) | Ar, He, CO_2, mixes (GTAW and GMAW) |

Figure 2.10 Forms of shielding for arc welding.

Welding fluxes serve other roles as well, including the very important function of stabilizing the arc by improving electron emission at the negative electrode. Elements that are referred to as deoxidizers or scavengers are added to remove undesirable materials such as oxides before they affect the weld. For example, such elements can make it possible to weld on steel with surface rust or mill scale. Other elements may be added to affect surface tension and improve the fluidity of the puddle. Iron powder is sometimes added to increase deposition rates, and alloying elements may be added to improve mechanical properties and form certain desirable metallurgical phases. Slag-forming elements produce molten slags that not only protect as described above but can also help shape the weld and assist in out-of-position welding. In the case of SAW, the flux that melts and floats to the top of the weld puddle plays a major role in the final shape of the weld bead. Table 2.1 shows some of the common elements that are added to create the various fluxes, and the roles that these elements play.

Table 2.1 Welding flux ingredients and their functions.

Ingredient	Function
Iron oxide	Slag former, arc stabilizer
Titanium oxide	Slag former, arc stabilizer
Magnesium oxide	Fluxing agent
Calcium fluoride	Slag former, arc stabilizer
Potassium silicate	Slag binder, fluxing agent
Other silicates	Slag binder, fluxing agent
Calcium carbonate	Gas former, arc stabilizer
Other carbonates	Gas former
Cellulose	Gas former
Ferro-manganese	Alloying, deoxidizer
Ferro-chromium	Alloying
Ferro-silicon	Deoxidizer

2.1.8 Weld Joints and Weld Types for Arc Welding

The selection of a proper weld joint and weld type is an important aspect of arc welding. Some very common arc welding joints and weld types are shown in Figure 2.11. The joint refers to how the two work pieces or parts that are being welded are arranged relative to each other. Weld type refers to how the weld is formed in the joint. Specifically in arc welding, there are numerous joint types, but only two weld types, namely, a fillet and a groove weld. A fillet weld (top two in the figure) offers the advantage of requiring no special joint preparation because the parts to be welded come together at an angle to form the necessary features to contain the weld. The strength of a fillet weld is a function of its size, as measured from the surface of the weld to the root of the weld where the parts meet.

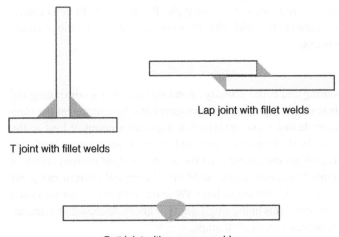

T joint with fillet welds

Lap joint with fillet welds

Butt joint with a groove weld

Figure 2.11 Typical arc welding joint and weld types.

Groove welds (bottom of figure) facilitate the creation of full penetration welds in thicker materials, which are usually necessary to generate maximum joint strength. The choice of weld and joint type is often dictated by the design details of the component being welded, and both often play a major role in the mechanical properties of the welded joint. The thickness of the parts being welded, as well as the material and type of welding process being used, also affects the choice of weld or joint type. Cost of welding almost always plays a role and is affected by the amount of preparation that is required to create the joint, the cost of filler metal, and how fast the weld metal can be deposited. Joints such as T and edge joints require minimal or no preparation since they only require the edges of the parts to be brought together. In the case of groove welds, the edges where the parts meet usually must be machined, ground, or thermally cut to produce the proper groove (although the figure shows a tight butt joint configuration in which special preparation may not be required). These subjects will be covered in more detail in Chapter 7.

Arc welded joints may be completed in a single weld pass or may require multiple passes. Single-pass full penetration groove welds in steel for instance are usually limited to about ¼ in. thick material or less. One of the advantages of all arc welding processes is that with multiple pass welding, materials of unlimited thickness can be created. In cases of extremely thick cross-sections, more than 100 weld passes might be required to fill the joint. When multiple pass full penetration welds are produced, the first pass, known as the root pass, is the most difficult. Subsequent passes are known as fill passes, and the passes which form the top of the weld are known as cap passes.

2.1.9   Primary Operating Variables in Arc Welding

2.1.9.1   Voltage
As mentioned previously, voltage is a variable that is important mainly for establishing a stable arc and is directly related to arc length. With the manual processes of SMAW and GTAW, the power supply maintains a nearly constant current, but the welder can affect the arc voltage by changing the arc length. With semiautomatic or automatic processes that use a continuous wire feed system, the arc voltage is selected on the power supply and will determine the arc length. In general, since the arc flares outward from the tip of the electrode to the flatter work piece, shorter arc lengths and lower voltages will tend to concentrate the arc heating in a smaller area while longer arcs spread the arc heating. Arc voltages with the GTAW process tend to be lower (10–20 volts), whereas consumable electrode arcs are more typically in the 15–40 volts range. Arc voltages also play a role in the shape of the weld with all processes, and in the metal transfer mode types with the GMAW process.

2.1.9.2   Current
Current plays a major role in arc heating and is the primary operating variable for controlling the amount of melting. Higher currents result in higher arc temperatures and heat inputs at the electrodes, melting the consumable electrode faster and transferring a greater amount of heat to the work piece creating a larger weld pool. With the manual arc welding processes, current levels are set at the machine, although some GTAW arrangements offer the ability to adjust current via a foot pedal allowing the welder to make modifications during welding. The typical current range for most arc welding processes is between 50 and 500 amps, but PAW and GTAW are sometimes used for very thin-walled applications at current levels in the single digits or lower. At the other extreme, SAW currents can approach and sometimes exceed 2000 amps.

2.1.9.3 Electrode Feed Rate/Wire Feed Speed

For SMAW, the welder feeds the electrode manually at a rate determined by the current setting on the power supply. If the welder is not feeding the electrode fast enough relative to the melt-off rate, then the arc length will lengthen, or the opposite happens if the feed rate is too fast. With the manual process of SMAW, electrode feed rate is not a measurable or predetermined variable, but something the welder must maintain relative to the electrode melting rate to keep a stable arc length.

When using the semiautomatic and automatic arc welding processes, electrode feed rate (or wire feed speed) is a setting on the power supply that determines both the electrode melting rate and the current. Typical wire feed speeds range between 100 and 500 inches/min. The wire feed mechanism uses an electric motor and a set of drive rolls that pull the electrode from the spool and push it toward the weld gun. Figure 2.12 shows a typical drive mechanism on a GMAW machine. Higher feed rates increase electrode melting rates and weld metal deposition. This is achieved due to the power supply automatically increasing current in response to the increase in wire feed speed. The change of current as wire feed speed is changed results in self-regulation of the arc length, an important factor with semiautomatic and automatic processes which utilize constant voltage power supplies. Therefore, with these processes, the setting of wire feed speed on the power supply determines the current. Since the self-regulation process of arc length is not perfect, a small adjustment of arc voltage is generally required to maintain the optimum arc length when the wire feed speed is changed. Modern power supplies that are called synergic automatically adjust the arc voltage with wire feed speed changes to keep the arc length constant.

Figure 2.12 Gas metal arc welding wire feed mechanism.

2.1.9.4 Welding Travel Speed

Welding travel speed refers to the rate at which the welding arc is moved along the joint. The heat input equation clearly shows that travel speed, like current, plays a direct role in the amount of heating into the part per unit length of weld. Faster speeds produce less heat into the work piece per unit length and reduced weld size. The choice of travel speed is typically driven by productivity;

faster welding speeds increase productivity and keep costs low. Travel speed is independent of current and voltage and may be controlled by the welder or mechanized. Typical travel speeds range between 3 and 100 inches/min.

2.1.10 Metal Transfer Mode

Filler metal transfer modes (Figure 2.13 shows one example) across the arc are often an important consideration with GMAW. The type of metal transfer mode can influence weld shape, heat input and depth of fusion, welding speed, spatter, and the ability to weld in different positions. The various transfer modes of spray or pulsed spray, globular, and short circuit are affected by many factors including wire diameter, current, voltage, and shielding gas. Spray transfer, illustrated in the figure, is a common mode in production because it provides for deep penetration, fast welding speeds, and minimal spatter. Another mode known as short circuit transfer occurs as the electrode periodically shorts into the weld pool. Historically, this mode has shown capability for welding thin materials but was known to be susceptible to producing lack of fusion defects. But as will be discussed later, many modern power supplies have capabilities for improved approaches to the short circuit transfer mode which offers the advantage of much lower heat input. As a result, this mode is growing in interest and frequency of application. Metal transfer modes are also discussed in much more detail later in the section on GMAW.

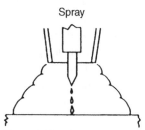

Figure 2.13 Spray transfer is one type of gas metal arc welding metal transfer modes.

2.1.11 Arc Blow

Arc blow is a phenomenon that can occur during arc welding, and results in the arc being deflected from the joint (Figure 2.14). The arc deflection is the result of a magnetic force known as the

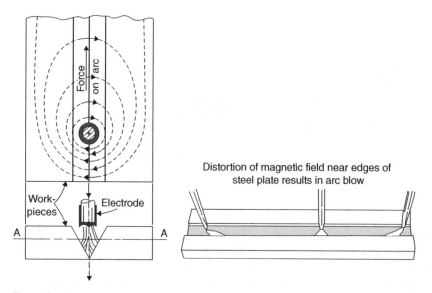

Figure 2.14 Arc blow. (*Source:* Reproduced by permission of American Welding Society, ©*Welding Handbook*).

Lorentz force. A Lorentz force occurs when a magnetic field interacts with a current-carrying conductor causing it to deflect. A magnetic field always surrounds a current-carrying conductor, and normally uniformly squeezes the conductor, so no net force deflecting force occurs.

In some cases, the magnetic field can become distorted around the arc, such as when welding near the edge of a steel plate as depicted in the figure. In this case, the ferromagnetic steel plate provides an easy flux path for the magnetic field which can result in a concentration of the magnetic field near the edge of the plate. This is because it is much easier for the magnetic flux lines to travel through the plate than through air, so even near the plate edge the preferred flux path remains in the plate. The resulting concentration of flux lines creates a greater magnetic Lorentz force on that side of the arc, pushing the arc in the opposite direction. This is just one form of arc blow, but the principles are always the same. Residual magnetic fields in a material can also deflect the arc in a less predictable manner.

2.1.12 Common Arc Welding Defects and Discontinuities

There are many possible forms of arc welding defects and discontinuities (or flaws). Some are metallurgical, while others are due to improper welding techniques. Common examples that arise from improper welding techniques include undercut, overlap, slag inclusions, and porosity. Figure 2.15 shows the undercut and overlap defects. Slag inclusions may also be the result of an improper welding technique, or in the case of multipass welds, insufficient cleaning between passes. Slag inclusions are possible only when welding with processes that use a flux, so weldments produced from processes such as GMAW, GTAW, and PAW are not susceptible to this defect. Porosity is a potential occurrence with all fusion welding processes and occurs when gasses such as hydrogen come out of the solution in the weld pool and form bubbles which get trapped as the molten metal cools and solidifies. It is particularly a problem when welding aluminum. The most common cause of porosity is improper cleaning of the work piece prior to welding. Weld defects and discontinuities (a weld "imperfection" or "flaw" that is not necessarily a defect) are covered in much more detail in Chapter 13.

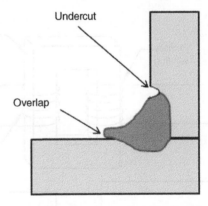

Figure 2.15 Undercut and overlap defects.

2.2 Arc Welding Power Supplies

The purpose of arc welding power supplies is to deliver the appropriate type of electrical power that is both safe and capable of producing a stable arc that can controllably melt the electrode and the metal being welded. Power sources come in a wide variety of configurations, from basic to very complex. At the most basic level, the power source must provide simple controls to allow the welder to adjust either the voltage or current output. More advanced arc welding power sources have features such as pulsing capability with preprogrammed pulse schedules customized for welding different materials and thicknesses. Other special features include the unique control of metal transfer modes, and interfaces designed to communicate with other equipment such as robots.

An arc welding power source produces arc welding electrical power in two ways—it either transforms it from incoming utility line voltage or generates it using an engine-powered generator. The transformer-type power sources are the most common. They receive the utility line voltage, which typically ranges from 240 to 480 volts or more in the US, and then transform it to relatively safe and usable open circuit arc welding voltages. As mentioned previously, 60–80 volts are common open circuit voltages for arc welding power supplies. The portable generator power supplies do not transform utility power, but directly generate the required welding voltages.

2.2.1 Transformers

Arc welding voltages are always much lower than utility voltages, so the most important function to condition utility power for arc welding is to reduce the voltage. A welding transformer is designed to receive high AC utility voltage on the primary or input side of the transformer and convert it to much lower welding voltages on the secondary or output side. These types of transformers are known as "step-down" transformers because voltage is reduced from the primary to the secondary. The top of Figure 2.16 reveals simple transformer construction which consists of primary and secondary windings wrapped around a common iron core. The bottom of the figure is the electrical symbol for a transformer. Windings on the primary side of the transformer convert high voltage to a magnetic flux from current flowing through the primary windings (or coils). The magnetic flux is then carried by the ferromagnetic transformer core to the secondary windings that induces a voltage across the secondary according to the ratio of secondary turns to primary turns, known as the transformer ratio. The voltage reduction in the "step-down" transformer requires a transformer with a smaller number of secondary turns than primary turns. The secondary voltage produces the open circuit voltage level of the power supply.

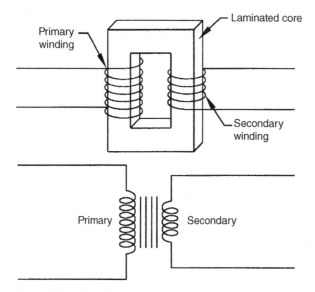

Figure 2.16 Transformer construction and electrical symbol. (*Source: Welding Essentials*, Second Edition).

2.2.2 Generators

For portability and use in the field, welding power supplies can be designed around engine-driven generators. A modern diesel-powered portable welding machine is shown on Figure 2.17. Generators work on the principle of a changing magnetic flux through conductive loops, which induce a voltage around the loops by Faraday's law of induction. Figure 2.18 shows a schematic of a DC generator with

Figure 2.17 Modern engine-driven welder (©Miller Electric Mfg. LLC).

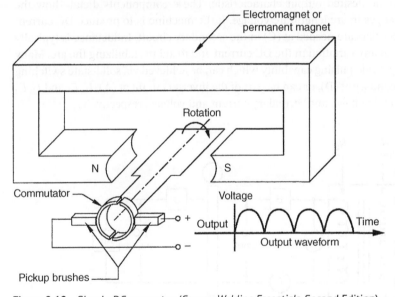

Figure 2.18 Simple DC generator. (*Source: Welding Essentials*, Second Edition).

one armature loop for simplicity. In practice, multiple loops or windings form the armature of the generator such that the output voltage of the connected loops sums to produce the total output voltage. In the DC generator, the armature loop is mounted on a shaft and is rotated by an engine in the gap between the magnetic poles. The resulting magnetic flux through the armature varies as its angle with the magnetic field changes inducing the output voltage. The voltage is in the form of an AC sinusoid at a frequency equal to the rotation speed of the armature in rotations per second. The AC voltage is then delivered through a pair of circular slip rings, one connected to each side of the armature via brushes that are made of graphite that slide on the rings. DC generators use a single split ring called a commutator (as shown in the figure) that causes the output voltage to always be of one polarity or DC.

An AC voltage can be produced more efficiently by a design in which the magnetic field member rotates instead of the armature. In this case, the magnetic field member is placed on a shaft inside the fixed armature. This configuration is called an AC alternator, or just an alternator. A drawback of the alternator is that mechanical commutation of the output to obtain DC is not possible since the armature is fixed. However, the use of modern solid-state rectifiers at the output can be used to convert to DC, representing the modern design for welding "generators." It should be noted that a much older technology for producing DC voltage prior to the invention of solid-state rectifiers was to use an AC motor to turn a DC generator. This older welding technology has largely disappeared.

2.2.3 Important Electrical Elements in Arc Welding Power Supplies

Arc welding can be performed with a very simple transformer or generator design as covered in the preceding section. However, there are many more elements that go into the design of modern welding power supplies. Figure 2.19 generalizes the location of important elements of a modern transformer-type power supply. On the primary side of the transformer at location A, advanced inverter-type power supplies use solid-state switching devices that convert the incoming line frequency to a higher frequency. The reason for this will be explained later in this chapter. The secondary circuit uses electrical and electronic components at B and D to adapt the secondary or welding outputs to give the desired output characteristic. These components dictate how the machine responds to changes in arc length/arc voltage. If the machine is to produce DC current (which almost all do), rectifiers are needed at (C) in the secondary circuit. Inductance is typically added at (D) to smooth out any variation in the DC current and to aid in stabilizing the arc. Many modern power supplies provide pulsing capability which can be achieved via solid-state switching devices located in the secondary at (D), or can be part of the inverter circuit at (A). I_1, E_1 and I_2, E_2 refer to primary current and voltage, and secondary current and voltage, respectively.

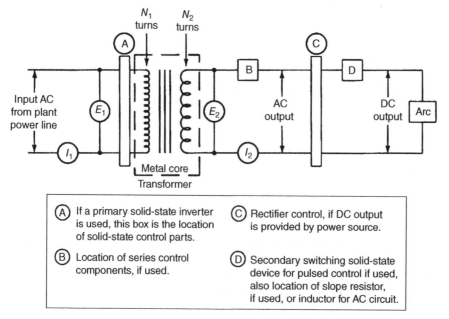

Figure 2.19 Typical elements of a modern arc welding power supply. (*Source:* Reproduced by permission of American Welding Society, ©*Welding Handbook*).

Power supplies may operate off single-phase (Figure 2.19) or three-phase utility power. Three-phase machines are widely used in the industry as they are more efficient, produce balanced loading on incoming power lines, and when converted to DC, result in less variation in the secondary wave form producing a smoother arc. Less expensive power supplies used in small shops or for farm and home use are usually single-phase.

What follows will provide a little more detail as to how modern electrical components can be used to design power supplies as generally depicted in the previous figure. Figure 2.20 shows a schematic of a typical rectifier arrangement in the transformer secondary to produce a DC output. A power supply of this type is often referred to as a rectifier. The rectifiers, arranged in a "full-wave bridge" configuration, convert the AC voltage in the secondary to DC by working together to allow current to only travel in one direction. During the AC half cycle when side A is positive, current travels in a loop that takes it through SCR1 and SCR4, and when side B is positive, current travels in a loop through SCR2 and SCR3. This configuration results in current that only travels in one direction, known as DC current. The inductance (Z) minimizes any variation in the waveform and stabilizes the arc.

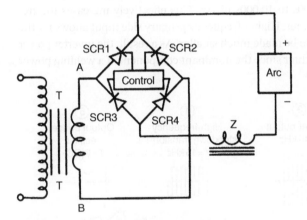

Figure 2.20 A typical rectifier bridge which produces DC current. (*Source:* Reproduced by permission of American Welding Society, ©*Welding Handbook*).

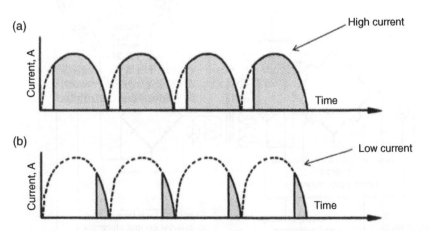

Figure 2.21 Phase control of current. (a) High-power conduction of SCR early in each half cycle and (b) lower-power conduction of SCR late in each half cycle. (*Source:* Reproduced by permission of American Welding Society, ©*Welding Handbook*).

The Silicon-Controlled Rectifiers (SCRs) of the above figure are rectifiers whose conduction can be controlled by a gate signal that is provided by an electronic controller. The controller provides what is known as phase control and serves to control the amount of current. With phase control, the controller allows current conduction through the SCRs at different times during the AC half cycle. With reference to Figure 2.21, to produce high output, the SCRS are turned on early in the half cycle (a) so they conduct a larger percentage of the time. To lower the output, the SCRs are turned on later in the cycle (b), and therefore, conduct a smaller percentage of the time.

Inverter power supplies have become very popular, primarily because of their light weight, small size, and advanced output controls. A typical inverter power supply may be 70% lighter or more compared to a conventional power supply design. The schematic of Figure 2.22 shows the basic electrical concepts of an inverter power supply, which relies on the generation of higher frequency alternating waveform for input to the primary side of the transformer. To accomplish this, the incoming 60 Hz utility line voltage is first rectified and converted to DC at location (2) shown in the figure. A series of solid-state switches then converts the DC current to a higher frequency at location (3), typically in the range of 1000s to 10,000s of Hz. This effectively increases the frequency on the primary side of the transformer. Higher frequency primary side input allows for the magnetic components in the transformer to be made much smaller. As a result, the inverter power supplies can be built much smaller and lighter since the dominant component of a welding power

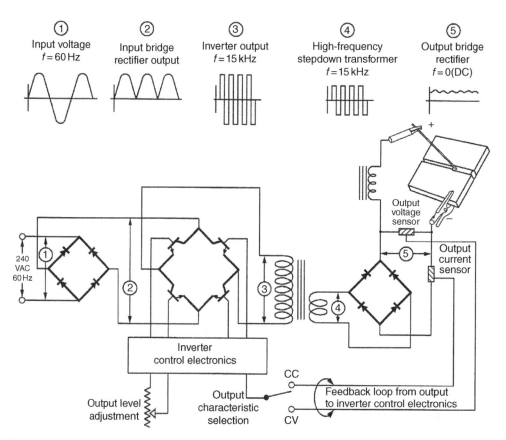

Figure 2.22 Basic electrical concepts of an inverter power supply. (*Source: Welding Essentials*, Second Edition).

supply is its transformer. Inverter power sources are more efficient, offer precise and rapid output control, and in many cases, produce a smoother arc due to minimal variation in the welding output. Feedback is an important element of these power supply designs as depicted in the lower right of the figure. The response to feedback from the sensing of arc voltage and current can be very fast since it occurs in the high frequency inverter section of the circuit.

Pulsing of the power supply output has become very popular, especially for GMAW, as well as some applications of GTAW. A typical pulsed DC current wave for GMAW is shown in Figure 2.23. The emergence of pulsing has arisen primarily due to the development of invertor technology (just discussed) that allows for much higher output control speed. Pulsing is typically in the range of 10s to 100s of pulses per second, depending on the application. Pulsed DC is advantageous for reduced heat input into the part while maintaining a good depth of fusion or penetration. It can also provide enhanced control of metal transfer with GMAW, especially in the maintenance of spray transfer. The hashed portion in the figure represents a typical transition current to spray transfer mode (this concept will be discussed later in Section 2.6). Pulsing can be designed to produce spray transfer only during these pulses, and thus reduce the average current level required for normal DC spray transfer. Many modern supplies come equipped with a wide variety of pulsing schedules, each designed and customized for specific combinations of base metal and filler metal.

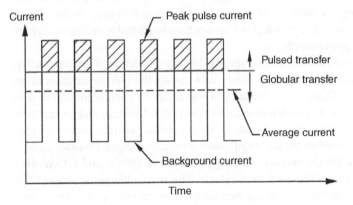

Figure 2.23 DC pulsed current signal. (*Source: Welding Essentials*, Second Edition).

2.2.4   Volt-Ampere Characteristic of Arc Welding Power Supplies

A defining feature of a welding power supply is its volt-ampere or V-A characteristic. A common type of V-A characteristic is illustrated in Figure 2.24 (arc voltages shown are typical for the GTAW process). The V-A characteristic shows how the power supply output voltage varies with output current and indicates whether the power supply is considered a constant voltage (CV) or a constant current (CC) type. The figure represents a CC-type characteristic. The V-A characteristic of a power supply can be developed experimentally by measuring and plotting its output voltage and current with various electrical loads (resistances). Since the typical welding power source delivers electrical power in the kilowatts to tens of kilowatts range, special test loads that can dissipate this level of heat have to be used. The first measurement of a V-A characteristic is taken with an open circuit. This corresponds to no load and zero output current. The open circuit voltage is 80 volts in the figure. The open circuit voltage is the voltage present after turning the power supply on and prior to striking an arc. The final data point can be taken

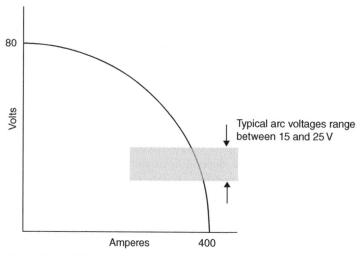

Figure 2.24 Volt-ampere characteristic of a constant current machine.

with no resistance at all, that is with the output short-circuited. This gives what is known as the short circuit current of the machine and represents the current that would flow if the welding electrode was shorted to the work during welding. The short circuit current is 400 amps in the figure. By introducing various loads, intermediate data points of current and voltage can then be plotted to produce the entire volt-ampere characteristic for the power supply.

As discussed in the previous sections, the arc voltage is primarily a function of arc length. The shaded region of the V-A characteristic shown in Figure 2.24 reveals the typical variation in arc voltage over a range of arc lengths. This is where the arc will operate under normal conditions with a CC power supply, where voltages are much less than the open circuit voltage and currents are somewhat less than the short circuit current. As can be seen, constant current power supplies allow only small changes in current when the arc length and arc voltage changes. Constant current machines are generally required for the manual welding processes of SMAW and GTAW since small changes in arc length during welding are unavoidable. The small changes in current that result from changes in arc length and arc voltage can provide for more welder control. Since these power supplies keep the current nearly constant, they are called constant current or CC machines.

The characteristic of a constant voltage (CV) power supply is illustrated in Figure 2.25 with the same arc voltage range shown in Figure 2.24. Notice that the open circuit voltage is much lower, and the short circuit current is much higher than for CC machines. Also, the much flatter characteristic as compared to CC machines means that small changes in arc length will produce very large changes in current. For this reason, CV power supplies are not usable for the manual processes of SMAW and GTAW. But constant voltage machines are ideal for semiautomatic and automatic processes where the electrode is fed continuously at a steady speed, referred to as the wire feed speed. CV power supplies provide for a process known as self-regulation of the arc length which maintains a stable arc length while the wire is being automatically fed. This process is explained on Figure 2.25. With reference to the figure, assume that the arc voltage (set on the machine) is at point B giving a current of 200 amps, which is just the right amount of current needed to melt the electrode at the rate it is being fed. Now consider that the welder makes an abrupt movement to momentarily increase the arc length which increases the voltage to point A. The machine senses this increase in voltage and reacts by dropping the current to 100 amps. This

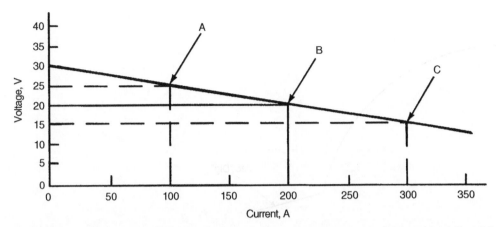

Figure 2.25 Volt-ampere characteristic of a constant voltage machine—note that locations A, B, and C show the large changes in current per small changes in arc voltage. (*Source:* Reproduced by permission of American Welding Society, ©*Welding Handbook*).

lowers the arc temperature which lowers the melting rate at the end of the electrode. Since the melting rate is reduced, but the wire feed speed remains the same, the arc length will automatically decrease, lowering the voltage back to the machine set point B. Similarly, if the welder abruptly decreases the arc length, the machine will respond by increasing the current which melts the electrode faster than the wire feed rate causing the arc length and voltage to increase and move back to B. This is the self-regulation of the process—arc length is a function of the arc voltage setting on the machine and is controlled by the machine and not the welder.

Because of self-regulation, semiautomatic and automatic welding processes are ideal for mechanized and automated welding since there is no need to control the arc length. Some slope to the V-A characteristic of the CV power supply is useful since it moderates the self-regulating process by reducing current fluctuations, making the process more stable. Inductance in the output can also help control sudden current surges, such as when the electrode shorts to the weld pool. The CV power supply with a constant wire feed system is used almost exclusively for GMAW, FCAW, and SAW.

Some modern power supplies can produce both constant voltage and constant current volt-ampere characteristics allowing them to operate a wide variety of arc welding processes. This is accomplished using electronic feedback control circuits to control the output of the welding machine, as was illustrated in the power supply circuit shown in the previous Figure 2.22. Such power supplies may also feature the capability to combine both characteristics on a single volt-ampere output curve (Figure 2.26). In this example, very low voltages due to a momentary short circuit cause the machine to act more like a constant voltage supply by delivering a high short circuit current. This capability of a power supply is sometimes called arc force or dig and can be helpful for preventing electrode sticking and providing for easier arc starting with the SMAW process. Once the arc is established or reestablished, the machine responds to the higher arc voltage and functions as a constant current machine.

The V-A characteristics shown represent the very basics of traditional welding power supply operation. Other modes of operation are possible and can be enhanced by modern electronic controls. For instance, there are methods for GMAW and SAW that use a wire feed speed adjustment for controlling arc length while using a constant current type of output. These systems rely on a

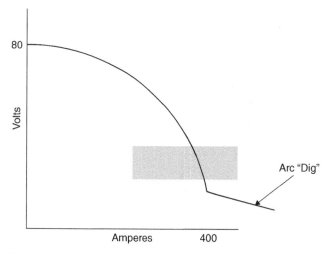

Figure 2.26 Power supply output provides both constant current and voltage, depending on arc length.

wire feed mechanism that can rapidly change speeds (and even direction) in response to arc voltage changes during welding. It should also be mentioned that the arc length in mechanized and automated GTAW can be controlled by an arc voltage sensing system and a controller that drives the weld torch up and down to keep the voltage and, thus, the arc length constant. These GTAW machines are called automatic voltage control (AVC) systems.

2.2.5 Duty Cycle

The so-called "duty cycle" of a welding power supply is an important consideration. Duty cycle refers to the maximum amount of time in a 10 min interval that a machine can be operated at its rated output. If a duty cycle is exceeded, the power supply is susceptible to overheating and will shut down before damage to electrical components occurs. The National Electrical Manufacturers Association (NEMA) provides standards for power supply duty cycles. NEMA has three categories of duty cycles referred to as classes. Class I machines have duty cycles between 60 and 100%, Class II between 30 and 50%, and Class III machines have a duty cycle of 20%. Class I machines are typically used in automated or high production environments where a power supply may operate for considerable lengths of time or even continuously for a given application. Class II machines are typically manual welding machines used in the industry where frequent starts and stops are common so that continuous operation is not required. Class III machines are for light duty where welding is very intermittent. Higher duty cycle machines require heavier electrical components and provisions for cooling, and thus are larger, heavier, and more expensive. On the other hand, Class III machines are small and less expensive.

Figure 2.27 depicts three Class I machines with three different rated outputs and a 60% duty cycle. The plot reveals that a machine rated at this duty cycle and rated current can operate at higher currents with a reduced duty cycle. Alternatively, machines can operate at longer duty cycles than their rated duty cycle with the operating current reduced below the rated current. For example, a 60% duty cycle machine rated at 300 amps can operate near 380 amps at a 35% duty cycle (3.5 min instead of 6 min out of 10 min). Welding power supplies always have a performance rating plate attached to the machine with further information provided in its operating manual.

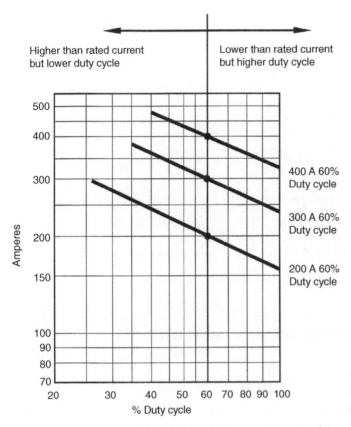

Higher than rated current but lower duty cycle

Lower than rated current but higher duty cycle

400 A 60% Duty cycle

300 A 60% Duty cycle

200 A 60% Duty cycle

Amperes

% Duty cycle

Figure 2.27 Class I machines with three different rated outputs. (*Source: Welding Essentials*, Second Edition).

2.2.6 Modern Advanced Arc Welding Power Supplies

Modern power supply manufactures produce a wide range of power supplies with a wide range of features. The most expensive machines offer high levels of sophistication and capabilities. This section will highlight some of the most advanced features of these power supplies.

Many modern GMAW power supplies offer significant advancements in the ability to weld utilizing the short circuit metal transfer mode. This mode allows for much lower heat input welding, and therefore, the ability to weld thinner material and root passes in multipass welds. As mentioned previously, although short circuit transfer mode GMAW has been available for a long time, it was known to be susceptible to welding problems such as lack of fusion and high amounts of spatter. But power supply manufacturers have made significant improvements with this transfer mode effectively eliminating these welding concerns.

Probably the most important attribute of these improvements in short circuit metal transfer is the ability to weld root passes in multipass welds, especially when welding pipe. Typically, root passes are welded with much slower SMAW or GTAW processes, so the ability to weld a root pass with GMAW offers the potential to significantly reduce welding time and cost. The approach to short circuit GMAW varies among the different manufacturers of power supplies. Terminology used by manufacturers for this form of metal transfer varies as well. Examples of terminology used include "Regulated Metal Deposition (RMD)," "Surface Tension Transfer (STT)," and "Cold Metal Transfer (CMT)." Figure 2.28 provides some detail to the RMD short circuit transfer mode. As the figure shows, a stable short circuit transfer mode is achieved by making precise adjustments to weld current to allow the wire to form a ball on the end and then dip (short) into the weld puddle.

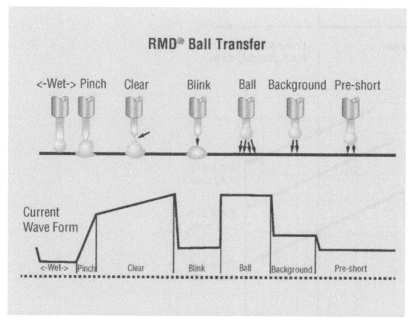

Figure 2.28 RMD short circuit metal transfer mode (©Miller Electric Mfg. LLC).

Figure 2.29 Advanced multiple process arc welding machine (©Miller Electric Mfg. LLC).

The most advanced machines offer multiple processes on a single machine and are specifically designed for pipe welding (Figures 2.29 and 2.30). A single machine can be used to weld with the SMAW, GTAW, GMAW, and FCAW processes, offer pulsing capability, and extremely easy set up as well. Some advanced modern power supplies offer the ability to produce spatter-free pulsed GMAW welds through sophisticated control of the waveform, including the ability to change the current ramp-up and ramp-down rate and shape (Figure 2.31 – Lincoln Electric's Power Wave Rapid X^{TM} Waveform Control Technology^{TM}).

Figure 2.30 Advanced multiple process arc welding machine (*Image provided by The Lincoln Electric Company, Cleveland, OH, USA*).

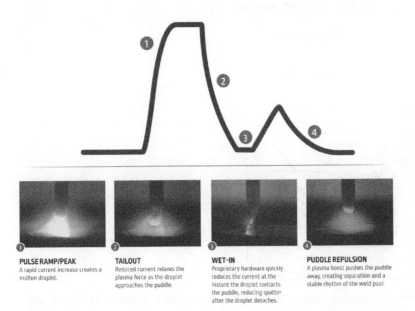

PULSE RAMP/PEAK
A rapid current increase creates a molten droplet.

TAILOUT
Reduced current relaxes the plasma force as the droplet approaches the puddle.

WET-IN
Proprietary hardware quickly reduces the current at the instant the droplet contacts the puddle, reducing spatter after the droplet detaches.

PUDDLE REPULSION
A plasma boost pushes the puddle away, creating separation and a stable rhythm of the weld pool.

Figure 2.31 Lincoln Electric's Power Wave Rapid X^{TM} Waveform Control Technology^{TM} allows for precision control of the arc for high-speed, ultra-low spatter for thinner material, with high-speed welding requirements (Reproduced with permission from Lincoln Electric's Power Wave® Rapid X^{TM} Waveform Control Technology^{TM}).

2.3 Shielded Metal Arc Welding

Shielded Metal Arc Welding (SMAW) is the most common arc welding process worldwide and has been in use since early in the twentieth century (Figure 2.32). It uses a covered electrode and requires no external shielding gas. The electrode covering is produced through an extrusion and baking process. It provides a variety of functions, but a primary function is to decompose when exposed to the arc heat to form CO_2 gas that protects the weld as it solidifies and cools. The electrode may also produce a protective slag that floats to the top of the weld puddle and solidifies. While other arc welding processes such as GMAW and SAW offer higher productivity, SMAW is known for the best versatility for use in both the shop and for field fabrication. The power supplies are relatively inexpensive, portable, and tend to be quite robust. It is commonly referred to as "Stick" Welding.

One of the many disadvantages of this process is the requirement for high welder skill level. Productivity is low due to a combination of slow welding speeds and the need to frequently stop and restart as the electrodes are consumed and replaced. The electrodes also produce high levels of welding fumes that can be a health hazard to both the welder and surrounding personnel, particularly in shop welding. Defect levels in SMAW deposits can also be quite high if proper techniques are not followed. Examples are porosity, incomplete fusion, slag inclusions, and poor bead shape. The chance for defects is higher in a stop and start location, so a weldment with many stops and starts is more prone to defects. In most cases, it is necessary to remove the slag following welding by wire brushing.

SMAW can be used with all common metals except for reactive metals such as titanium, which is extremely sensitive to interstitial embrittlement and requires inert gas shielding. It is possible

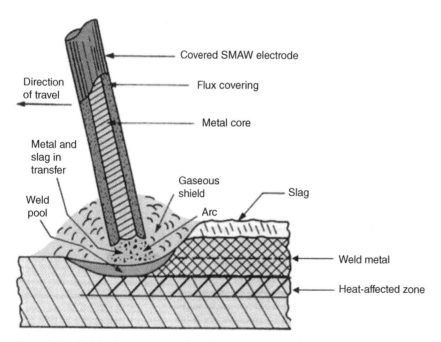

Figure 2.32 Shielded metal arc welding. (*Source:* Reproduced by permission of American Welding Society, ©*Welding Handbook*).

to weld aluminum, but the applications are limited. Generally, the minimum plate thickness when welding with SMAW is 1/8 in. thick. While it is possible to weld thinner plates, great welder skill is required to avoid melting through the part. Because of the ability to produce multipass welds, there is no limit to the maximum plate thickness that can be welded with this process. However, with large plate thicknesses (and therefore, large numbers of passes required), SMAW becomes less economical and more likely to be replaced with a higher deposition process such as FCAW or SAW. SMAW can be used in all positions, but not all electrodes can be used in all positions.

As mentioned previously, in addition to decomposing to form a protective gas (CO_2) cover to shield the weld metal as it solidifies, the SMAW electrode coating may provide a wide variety of other functions. Scavengers and deoxidizers are added to allow for welding on metals that are not clean and/or contain rust, scale, and other oxides. Alloy elements can be added to the electrode coating to produce desirable weld metal microstructures and improve the mechanical properties of the weld metal. Iron powder is sometimes added to improve the deposition rate. The slag that is produced by some electrodes may not only provide additional protection from the atmosphere, but can play a role in enhancing the bead shape.

Figure 2.33 shows the typical effects of welding amperage, arc length, and travel speed on the visual appearance of the weld. These effects may vary depending on the type of metal being welded, the type and size of the electrode, and the polarity being used. SMAW usually uses the DCEP polarity which produces the best depth of penetration. But some electrodes are designed for DCEN which results in less part heating and greater electrode melting rates. Many electrodes can also be used with AC current, but DC current will usually provide a more stable arc. Arc currents typically range between 100 and 300 amps, but may be higher or lower than this range depending on the electrode diameter and type. The amount of current directly affects the electrode melting rate and heating of the work piece. And as discussed previously in this chapter, travel speed also plays a direct role in heat input to the part. Slower travel speeds will produce larger welds and higher heat input and vice versa.

(A) Proper amperage, arc length, and travel speed

(B) Amperage too low

(C) Amperage too high

(D) Arc length too short

(E) Arc length too long

(F) Travel speed too slow

(G) Travel speed too fast

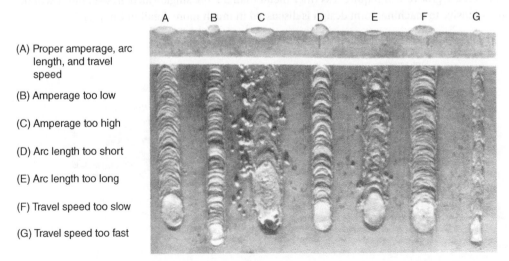

Figure 2.33 Typical effects of amperage, arc length, and travel speed. (*Source:* Reproduced by permission of American Welding Society, ©*Welding Handbook*).

A proper arc length is one that produces the smoothest arc and metal transfer. The arc length also establishes the arc voltage, which normally ranges between 20 and 30 volts. Excessive lengths will result in significant spatter and a flat weld profile and may also reduce the effectiveness of the gas shielding, thereby promoting metal contamination and porosity. Arc lengths that are too short may create short circuit transfer of the filler metal, thereby reducing heating and promoting spatter. A general rule of thumb is that the arc length should be equal to the diameter of the wire core in the electrode being used.

In addition to the variables of amperage, arc length, and travel speed, the orientation of the electrode relative to the work piece is also an important variable (Figure 2.34). There are two angles to consider—the work angle, which is the angle between the electrode and the work piece, and the travel angle, which is the angle of the electrode relative to the direction of travel. Another important consideration for the welder is whether to use a forehand or a backhand welding technique. As indicated in the figure, the electrode is pointed in the direction of travel when forehand welding is used, and in the opposite direction when welding with a backhand technique. Forehand and backhand angles are often referred to as "push" and "drag" angles respectively. Many factors will affect the choices of electrode angle and welding technique, including joint design, electrode type, and welding position. Table 2.2 lists typical work and travel angles that are used as a function of weld type and welding technique.

Joint design is a very important issue with all arc welding processes. Examples of typical designs used for SMAW are shown in Figure 2.35. Depending on the application, a joint may need to be designed to provide the ability to produce a full penetration weld. Full penetration with relatively thin parts may be possible with a simple gap between the parts, whereas thicker parts will require groove angles to be machined or cut. A groove angle that is too narrow may result in weld defects such as lack of fusion or slag inclusions. A groove angle that is too large may result in an excessive number of weld passes which will significantly increase the time and cost associated with producing the weldment. Joints must be designed that allow electrode access and the ability to properly fixture. Backing bars, which are usually removed after welding, may be used to assist in creating a full penetration weld, but do not allow for a visual inspection of the root pass. When deciding on a joint design, machining and cutting costs must be balanced against welding costs. For example, a double-sided V groove will require less filler metal than a large single-sided groove, but it will be more expensive to machine. Joint design is discussed in much more detail in Chapter 7.

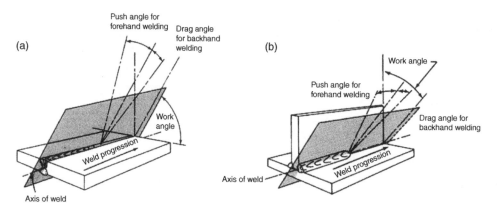

Figure 2.34 Electrode orientation and terminology. (a) Groove weld and (b) fillet weld. (*Source:* Reproduced by permission of American Welding Society, ©*Welding Handbook*).

Table 2.2 Typical shielded metal arc welding work and travel angles.

Type of joint	Position of welding	Work angle, deg	Travel angle, deg	Technique of welding
Groove	Flat	90	5–10[a]	Backhand
Groove	Horizontal	80–100	5–10	Backhand
Groove	Vertical—Up	90	5–10	Forehand
Groove	Overhead	90	5–10	Backhand
Fillet	Horizontal	45	5–10[a]	Backhand
Fillet	Vertical—Up	35–55	5–10	Forehand
Fillet	Overhead	30–45	5–10	Backhand

Source: Reproduced by permission of American Welding Society, ©*Welding Handbook*.
[a]Travel angle may be 10–30° for electrodes with heavy iron powder coatings.

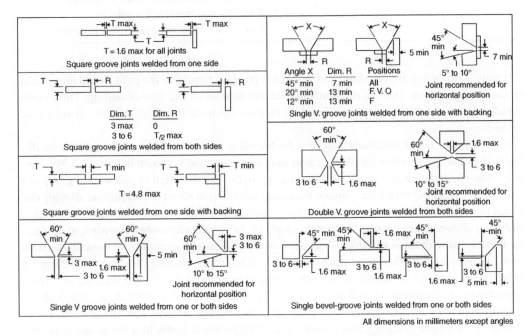

Figure 2.35 Examples of shielded metal arc welding joint designs. (*Source:* Reproduced by permission of American Welding Society, ©*Welding Handbook*).

AWS specifies electrodes for SMAW by the alloy family they are to be used for. The AWS standard that serves as an SMAW electrode specification for welding carbon steel is known as A5.1. In the five-digit designation system shown in Table 2.3, the E refers to electrode. This means that the filler material is also part of the electrical circuit and that it can only be used this way (i.e., this electrode cannot be used as filler wire for the GTAW process). The first two numbers indicate the minimum ultimate tensile strength (ksi) of the weld metal that will be produced by this electrode. The third digit refers to the weld position the electrode can be used for. This designation reflects the characteristics of the weld puddle that may affect out-of-position welding, such as weld puddle "fluidity" and surface tension. For example, the electrodes that only work in the flat and horizontal positions

Table 2.3 Electrode specification system for carbon steel electrodes.

EXXXX
- E = electrode
- First two or three digits = ultimate tensile strength (UTS) of weld metal in ksi
- Second from last digit = position
- Last digit = type of coating/current

Third digit	Fourth digit
• 1 all positions	0 DCEP, cellulose-sodium silicate
• 2 flat and horizontal	1 AC/DCEP, cellulose, potassium
• 4 flat, overhead, horizontal, vertical down	2 AC/DCEN, rutile, sodium
	3 AC/DC, rutile, potassium
	4 AC/DC, rutile/iron powder
	5 DCEP, lime, sodium, low hydrogen
	6 AC/DCEP, lime, potassium, low hydrogen
	7 AC/DC, iron oxide/iron powder
	8 AC/DCEP, lime/iron powder, low hydrogen

(2) create puddles that are very fluid and would not properly remain in the joint if used for vertical or overhead positions. The last digit provides information about the type of coating. The recommended polarity and current type (AC vs. DC) for each electrode is a function of the coating type as shown.

The AWS filler metal specification system for low-alloy steel electrodes (AWS A5.5) is a bit different than the specification for plain carbon steel electrodes and is shown in Table 2.4. The low-alloy steel specification includes a second series of digits. The first digit of this second group is a letter that may be followed by a number, providing information about the additional alloying elements that will make up the weld metal deposit. The next two digits refer to the maximum level of hydrogen allowed in the coating and consist of an "H" followed by a number. A hydrogen designator is provided with these electrodes because low-alloy steels will generally be more hardenable than carbon steels, and therefore, more susceptible to hydrogen cracking. Hydrogen cracking is a common and potentially catastrophic form of cracking associated with the welding of ferrous materials and is discussed in detail in Chapter 10. If the designation ends with an "R," the electrode is designated to be resistant to moisture pick-up during storage. Moisture in flux is a major contributor to hydrogen cracking, and therefore, keeping moisture levels to a minimum is often a major priority with any processes that rely on a flux.

One more example of an SMAW AWS electrode specification is A5.4, "Specification for Stainless Steel Electrodes for Shielded Metal Arc Welding." In this case, the base metal specification number is used followed by a two-digit number that provides information about the welding position, coating type, and the polarity the electrode is designed for. An example is "E308-15." The 308 refers to the stainless steel alloy type, the first number (1) refers to the welding position the electrode can be used for, and the last number (5) provides information about the coating type and polarity. When the welding position indicator is a 1, the electrode can be used in all positions if it is not larger than 5/32 in. diameter. If the electrode is larger than 5/32 in. diameter, or the welding position indicator is a 2, the electrode can only be used in either the flat or horizontal positions. When the last digit is a 5, the electrode is limestone based and suitable for DCEP only. Specifications that end with a 6 or 7 refer to electrode coatings that are based on titanium dioxide and can be used with either AC

Table 2.4 Electrode specification system for low-alloy steel electrodes.

	Chemical composition of weld deposit				
E 8018–B1H4 R Electrode ——— 80,000 PSI Min. —— All position ——— For AC or DCEP ——— Chemical composition of weld metal deposit Moisture resistant designator indicates the electrode's ability to meet specific low moisture pickup limits under controlled humidification tests Diffusible hydrogen designator indicates the maximum diffusible hydrogen level obtained with the product	Suffix	%Mn	%Ni	%Cr	%Mo	%V
	A1				1/2	
	B1			1/2	1/2	
	B2			1-1/4	1/2	
	B3			2-1/4	1	
	C1		2-1/2			
	C2		3-1/4			
	C3		1	0.15	0.35	
	D1 and D2	1.25–2.00			0.25–0.45	
	G[a]		0.50	0.30 min	0.20 min	0.10 min

[a] Only one of the listed elements required.

or DCEP polarity. These three examples provide clear examples of the differences in the various AWS SMAW electrode specifications. For the Welding Engineer who is going to be working with SMAW or any arc welding processes, it is important to obtain and become familiar with the applicable electrode specifications.

The different electrode coatings play a major role in the "usability" characteristics of the electrode. For example, some electrode coatings produce weld puddles and molten slag that are very "sluggish" while others produce very fluid puddles. Other differences produced by the various coatings include depth of penetration, shape of the weld (convex vs. concave), and the amount of spatter. The slag from some electrodes may be easier to remove than the others.

It is also true that electrodes with different specification numbers may have similar characteristics. For this reason, the ASME (American Society of Mechanical Engineers) Boiler and Pressure Vessel Code Section IX includes a categorization scheme based on usability characteristics of the electrode. This permits the welder to be able to change electrodes to a different type without the need to requalify the welding procedure, as long as the substitute electrode has the same usability characteristics. Qualifying a procedure is a time-consuming and expensive process often mandated by welding codes and is discussed in Chapter 14. For carbon (or mild) steel, ASME categorizes electrodes by "F" number. There are four groups—F-1 through F-4. The F-1 group electrodes produce the highest deposition rates of all the groups, but are limited to flat and horizontal positions. They are known for smooth, nearly ripple-free beads with minimal spatter. The F-2 group electrodes produce minimal penetration, and therefore, are excellent for welding thin plates. They are typically used with DCEN. The F-3 group electrodes produce a forceful arc and are known for deep penetration. They also solidify rapidly and therefore work very well in all positions. The forceful, deep penetrating arc combined with rapid solidifying capabilities make these electrodes ideal for root passes. They produce a light slag and include coating additions that make them ideal for welding dirty or painted material. The F-4 group is the low hydrogen group. These electrodes should be used when there is a possibility of hydrogen cracking.

In summary, the SMAW Process offers the following advantages and limitations:

Advantages:

- Inexpensive equipment
- Portable
- With the proper electrode, can weld in all positions
- Ability to weld in drafty conditions
- Tolerant to less than ideal joint fit-up
- Through multiple pass welding, there is no limit to the maximum thickness that can be welded

Limitations:

- Process is slow, primarily due to the need to frequently replace electrodes
- The discarded electrode stubs represent waste
- Requires relatively high welder skill
- Becomes difficult to use when plate thicknesses are less than 1/8 in.
- The frequent stops and starts to replace the electrodes increase the likelihood of a defect
- Inability to weld reactive metals such as titanium
- Electrodes are sensitive to moisture absorption and may require special storage to reduce the possibility of hydrogen cracking when welding certain steels

2.4 Gas Tungsten Arc Welding

Gas Tungsten Arc Welding (GTAW) is somewhat unique among arc welding processes in that it uses a nonconsumable tungsten electrode to establish the arc (Figure 2.36). Since the arc and filler wire are independent of each other, GTAW offers the possibility for much more precise control of heat input and the weld puddle. Because the process does not involve molten filler metal passing through the arc, there is no spatter. Combined with the fact that it usually uses an inert shielding gas, it is often considered to be capable of producing the highest quality welds of all the arc welding processes. In some cases, it also offers the option of welding without filler metal, known as autogenous welding. The process is usually referred to by its original name, "TIG" (Tungsten Inert Gas). It was first used in the United States in the 1940s when it was called Heliarc (because it used helium shielding gas), a name that is sometimes still used today.

The GTAW process is used for a wide variety of commercial applications. Since it can be used with very low levels of current and with no filler metal addition, it is often used for small components where heat buildup is a problem. The medical products and electronics industries use GTAW extensively to make final closure welds on sensitive products such as pacemakers and batteries. It can also be used for the manufacture of tubing, both for longitudinal seam welds and for connecting sections of tube (orbital tube-to-tube welds). It is also commonly used in the manufacture of heat exchangers where the tubes are connected to thick plates known as tube sheets.

GTAW is a popular process for a wide variety of repair welding since heat input can be precisely controlled and directed. For example, in the gas turbine engine industry, GTAW is used to repair blade tips, or small cracks or defects in turbine engine components made from stainless steel, titanium, or nickel-based alloys. Precise heat control also makes it an ideal process for producing the root pass of a joint between two pipes. Therefore, whenever high-quality welds and precise heat control are required, GTAW is often the process of choice. However, it is not the process of choice

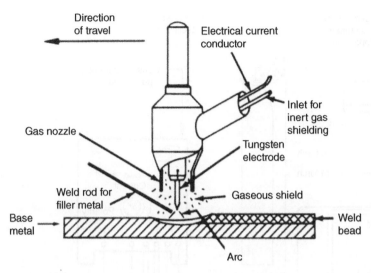

Figure 2.36 Gas tungsten arc welding. (*Source:* Reproduced by permission of American Welding Society, ©*Welding Handbook*).

when high productivity is paramount. Also, because the shielding comes from a gas being delivered from a nozzle, it is sensitive to welding in drafty conditions.

The typical GTAW equipment set-up (Figure 2.37) includes a constant current power supply and a torch (Figure 2.38) that may or may not use water cooling. Water cooling is needed mainly for high-current (typically 200 amps or greater) and high-duty cycle applications, which typically involve mechanized or automatic welding. Another unique feature of this process is the optional foot pedal that adds another dimension to welder control by providing for very precise adjustments of current as the weld is being made. GTAW power supplies can also be used for SMAW.

GTAW is limited to single-pass full penetration maximum weld plate thicknesses of about 1/8 in. Arc currents can be less than 1 ampsor as high as 500amps, but usually fall in the range of 30–150amps. Arc voltage ranges of 10–20 volts are typical, which is considerably less than other arc welding processes. Welding travel speeds are typically 3–6 inches/min. Autogenous welding is common when welding thinner materials or when welding along edges. Argon is the most common shielding gas primarily because it is less expensive than helium, and because it is heavier than air (and helium), it is not as sensitive to drafts, and lower flow rates can be used. Helium has the advantage of creating an arc with higher thermal conductivity, so it transfers heat more efficiently from the arc to the work piece. Argon and helium blends are often used to combine the benefits of each gas. Hydrogen is sometimes added to argon to increase penetration when welding stainless steel.

GTAW mostly utilizes DCEN polarity which provides the most heat into the part, and therefore, the deepest penetrating weld. Because this polarity also minimizes the heat into the electrode, electrode life is much better than with the other polarities. Since DCEP results in significant heating and melting of the electrode and produces a wide and shallow weld profile, it is rarely used. However, DCEP offers a significant benefit when welding aluminum known as cathodic cleaning of the aluminum oxide on the surface. As the electrons leave the surface and the ions bombard it, the oxide layer is removed. This is important when welding aluminum because the aluminum

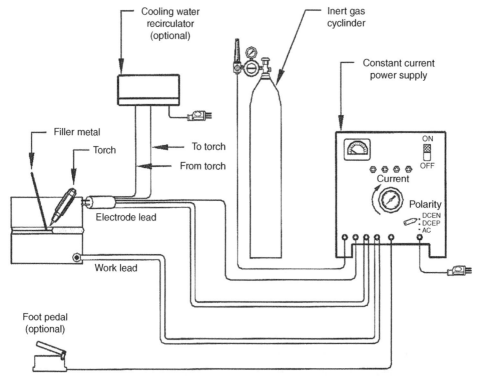

Figure 2.37 Gas tungsten arc welding equipment set up. (*Source: Welding Essentials*, Second Edition).

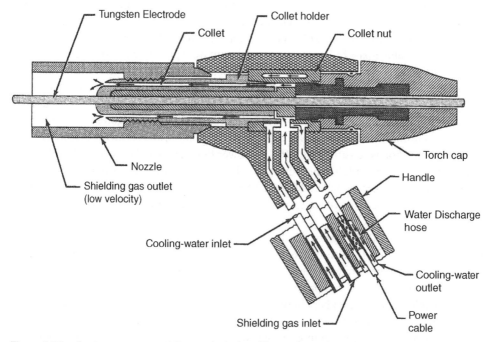

Figure 2.38 Gas tungsten arc welding torch design. (*Source:* Reproduced by permission of American Welding Society, ©*Welding Handbook*).

oxide significantly reduces the wetting action of the puddle. But due to the very high thermal conductivity of aluminum, DCEP cannot generate enough heat to counter the heat extraction capability of aluminum. Therefore, it is very common to use AC when welding aluminum because it provides half a cycle of oxide cleaning action and half a cycle of higher heat into the part. Advanced power supplies with pulsing capability provide the opportunity to customize the cleaning action and heating pulses when welding aluminum. The advantages and disadvantages of the various polarities discussed are summarized in Figure 2.39. Pulsed current can also be used with DCEN, typically to help control weld penetration when welding sheet metal. High pulse current is used to provide penetration into the base metal, and low pulse current allows the metal to partially solidify to prevent excessive melting.

Current type	DCEN	DCEP	AC (balanced)
Electrode polarity	Negative	Positive	
Electron and ion flow			
Penetration characteristics			
Oxide cleaning action	No	Yes	Yes–once every half cycle
Heat balance in the arc (approx.)	70% at work end 30% at electrode end	30% at work end 70% at electrode end	50% at work end 50% at electrode end
Penetration	Deep; narrow	Shallow; wide	Medium
Electrode capacity	Excellent e.g., 3.2 mm (1/8 in.) 400 A	Poor e.g., 6.4 mm (1/4 in.) 120 A	Good e.g., 3.2 mm (1/8 in.) 225 A

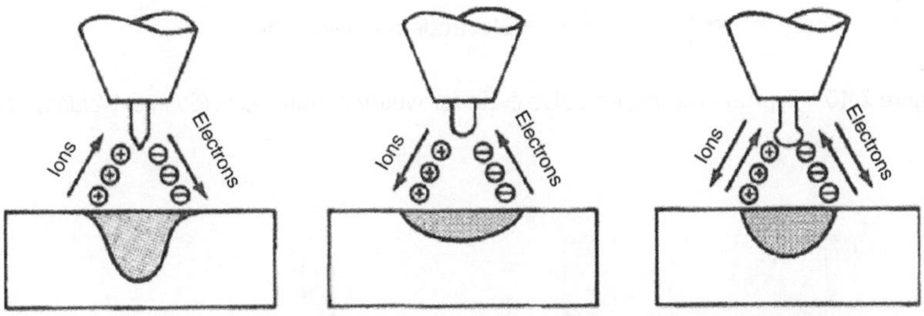

Figure 2.39 Effect of polarity with the gas tungsten arc welding process. (*Source:* Reproduced by permission of American Welding Society, ©*Welding Handbook*).

Figure 2.40 shows a "customized" pulse cycle that might be used for welding aluminum. The electrode negative pulse is relatively short in time but high in current. This provides good penetration with less overall heat into the part. The electrode positive pulses are wider but lower in peak current. This allows for good cleaning action without excessively heating the electrode. The pulsed wave form that includes positive and negative polarities can be expected to be more stable than a conventional AC waveform due to the vertical slopes on the waveform as the current changes direction. This means that there is less time when the current is near zero as compared to a conventional AC waveform.

To form a properly shaped arc, the tip of the tungsten electrode must be ground to the proper angle. A wide variety of grinding machines are available for this purpose, some of which allow for precise control of the tip angle. In general, smaller tip angles will result in a wider arc that is less dense, which in turn produces a wide weld profile with shallow penetration. Blunt tip angles produce a more concentrated arc with narrow and deeper penetrating weld profile (Figure 2.41). Once the choice of angle is determined for a given application and/or procedure, it is important to always maintain that angle. When welding with AC, it is common to form a ball on the electrode before welding. This can be accomplished by striking the arc on a piece of copper while using DCEP and

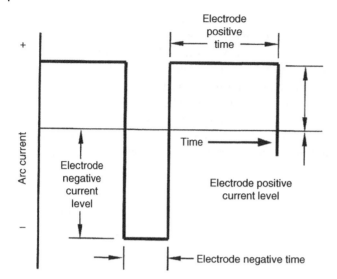

Figure 2.40 Typical customized pulse cycle for welding aluminum. (*Source: Welding Essentials*, Second Edition).

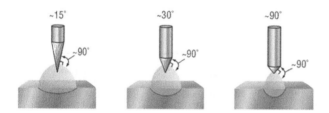

Figure 2.41 Effect of electrode tip angle on arc and weld profile (©Miller Electric Mfg. LLC).

increasing the current until a molten ball forms on the end of the electrode. The current can then be ramped down to allow the ball to solidify.

A common application of GTAW known as orbital welding is a mechanized weld process for high-quality welding of pipe sections. Orbital welding uses a GTAW torch on a mechanized weld head attached to the pipes (or tubes) along the joint to be welded. The torch then travels around the pipe joint along the weld head. Since the variables of arc length and travel speed can be precisely controlled, faster and more consistent pipe welds can be made as compared to manual welding. Orbital GTAW is not totally automated though; operation and close observation of the welding operation must be conducted by a qualified welder. Figure 2.42 shows a typical customized pulsing schedule typical for this type of welding. The current is gradually tapered down toward the end of the weld to reflect the fact that the pipe will get hotter as the welding torch progresses around it. If the current is not properly tapered down, burn-through of the pipe could occur.

A typical manual welding technique with GTAW (Figure 2.43) involves first forming a molten weld puddle and then establishing a forehand (or push) angle of about 15°. The filler metal (if needed) is then dipped into the leading edge of the puddle as the weld progresses. The dipping action continues throughout the weld; as the filler metal is dipped into the puddle, enough of the wire melts to properly shape the puddle. When the filler metal is lifted out of the puddle it is important for the welder to keep the hot end of the wire under the flow of shielding gas so that it does not get contaminated. And by keeping the angle of the filler wire as low as possible it is less likely that the wire will contact the tungsten electrode.

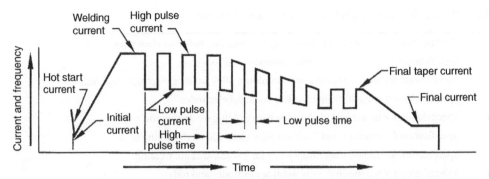

Figure 2.42 GTAW pulsing schedule designed to reduce heat near the end of the weld. (*Source: Welding Essentials*, Second Edition).

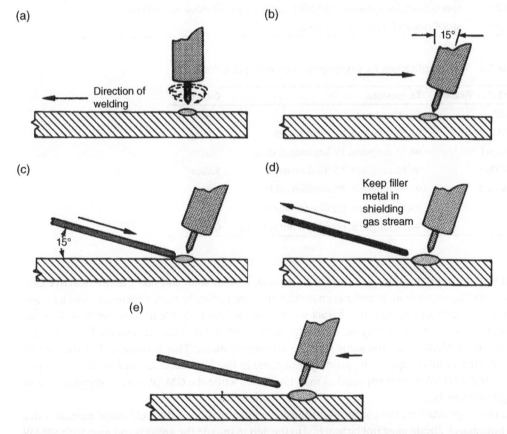

Figure 2.43 Typical gas tungsten arc welding technique. (a) Develop the pool with circular or side-to-side motion, (b) move electrode to trailing edge of pool, (c) add filler metal to center of leading edge of pool, (d) withdraw filler metal, and (e) move electrode to leading edge of pool. (*Source:* Reproduced by permission of American Welding Society, ©*Welding Handbook*).

Tables 2.5 and 2.6 list the AWS specifications for GTAW Filler Metal and Electrodes, respectively. The electrodes consist primarily of tungsten with the alloying additions shown. The alloying additions (oxides) improve thermionic emission of the tungsten, which allows the electrodes to be operated at higher currents. The additions also result in longer electrode life, easier arc starting, and a more stable arc. The thoriated electrodes have historically been the most common, but there are safety concerns with this electrode due to the fact that thorium is radioactive

Table 2.5 AWS specifications for gas tungsten arc welding filler metals.

A5.7	Specification for copper and copper alloy bare welding rods and electrodes
A5.9	Specification for bare stainless steel electrodes and rods
A5.10	Specification for bare aluminum and aluminum alloy welding electrodes and rods
A5.13	Specification for solid surfacing welding rods and electrodes
A5.14	Specification for nickel and nickel alloy bare welding electrodes and rods
A5.16	Specification for titanium and titanium alloy electrodes and welding rods
A5.18	Specification for carbon steel filler metals for gas shielded arc welding
A5.19	Specification for magnesium alloy welding electrodes and rods
A5.21	Specification for composite surfacing welding rods and electrodes
A5.24	Specification for zirconium and zirconium alloy welding electrodes and rods
A5.28	Specification for low-alloy steel filler metals for gas shielded arc welding
A5.30	Specifications for consumable inserts

Table 2.6 AWS specifications for gas tungsten arc welding electrodes.

AWS classification	Composition	Color code
EWP	Pure tungsten	Green
EWCe-2	97.3% tungsten, 2% cerium oxide	Orange
EWLa-1	98.3% tungsten, 1% lanthanum oxide	Black
EWTh-1	98.3% tungsten, 1% thorium oxide	Yellow
EWTh-2	97.3% tungsten, 2% thorium oxide	Red
EWZr-1	99.1% tungsten, 0.25% zirconium oxide	Brown
EWG	94.5% tungsten, remainder not specified	Gray

(although at a very low level). As a result, ceriated, lanthanated, and other nonradioactive electrodes are becoming more popular even though they are generally more expensive. Another issue with these electrodes is their arc characteristics may be less desirable to some welders. To minimize the chance of misidentifying an electrode, they are color coded as indicated on Table 2.6. GTAW and GMAW share the same filler metal specifications. This is discussed in the GMAW section later in this chapter. The primary difference in the filler material used for these two processes is the GTAW wire is supplied as individual rods while the GMAW wire is supplied in long lengths on spools.

In many applications, the concern for tungsten contamination of the weld metal mandates that the tungsten electrode must not be touched to the part to initiate the arc as is common with SMAW, so other arc starting approaches are needed. One approach is a pilot arc system inside the nozzle, but a more common approach is a high frequency arc starting system which produces an extremely high voltage and low amperage arc to get the ionization process started (Figure 2.44). These high frequency systems can adversely affect electronic equipment so care must be taken if any electronic equipment is nearby.

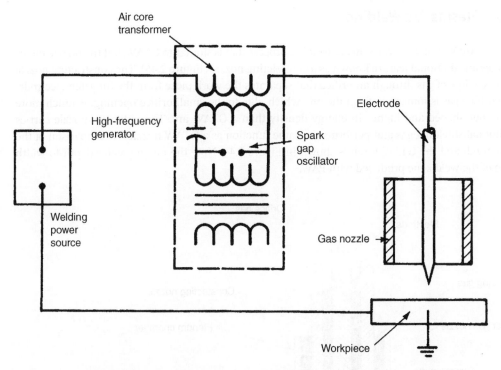

Figure 2.44 High frequency starting system for gas tungsten arc welding. (*Source:* Reproduced by permission of American Welding Society, ©*Welding Handbook*).

In summary, the GTAW Process offers the following advantages and limitations:

Advantages:

- Due to the precise heat and puddle control, the absence of spatter, and the inert gas shielding, it is generally known to produce the highest quality welds of all arc welding processes
- Since there is no flux used, there is minimal postweld cleaning required and no possibility for a slag defect
- Can weld all nearly all metals and be used in all positions
- Works well for complex geometries and thin sheets
- Excellent for root passes
- Provides the option for autogenous welding with some applications

Limitations:

- Requires very high welder skill
- Low weld metal deposition rates
- Sensitive to drafty conditions
- Possibility for tungsten inclusions in weld
- Arc starting system adds cost

2.5 Plasma Arc Welding

Plasma Arc Welding (PAW) is an arc welding process that is similar to GTAW, but the nozzle incorporates an additional feature known as a constricting nozzle (Figure 2.45). The constricting nozzle directs a flow of gas through an orifice that separates the work piece from the tungsten electrode. The orifice gas is ionized to form the arc, which, due to the small orifice opening, is much more columnar shaped and higher in energy density than a GTAW arc. The outer gas nozzle carries additional shielding gas and performs the same function as a GTAW nozzle. A comparison of the PAW torch and the GTAW torch is shown in Figure 2.46. Note the much greater depth-to-width ratio of the weld zone produced with PAW.

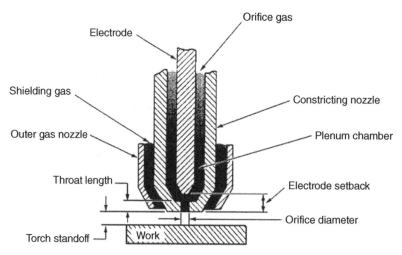

Figure 2.45 Plasma arc welding. (*Source:* Reproduced by permission of American Welding Society, ©*Welding Handbook*).

PAW was developed commercially by the Union Carbide Linde Division in the 1960s. It saw little use until the 1970s through 1990s, and is still not as widely used as competing processes, such as GTAW. Industry sectors where it is used include aerospace, automotive, medical, tube mills, and electrical. Filler metal may or may not be used, and the equipment is more expensive than other arc welding equipment. Because the electrode is contained within the constricting nozzle and is farther from the work piece it is less susceptible to contamination than the electrode used for GTAW. The electrodes and filler metals used are the same as those used with GTAW. Argon and argon–helium blends are the most common gasses for both the orifice gas and the shielding gas.

Figure 2.47 provides evidence of the advantage of arc constriction, which is depicted on the right side of the figure. A restricted arc contains a much hotter core that extends for a longer distance than a standard GTAW arc, which tends to rapidly flare out from the electrode losing energy density. The greater energy density and columnar shape of the constricted arc allows for keyhole mode welding (Figure 2.48), which can produce single-pass welds of much greater thickness, faster welding speeds, and lower heat input compared to GTAW. As indicated in the figure, this mode of welding involves the formation of a hole that usually traverses all the way through the joint. The molten weld metal swirls around the hole and solidifies at the trailing edge as the weld is moved along the joint. Although this mode of welding offers the advantages mentioned, it is very difficult to

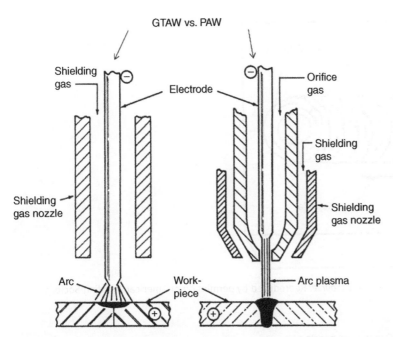

Figure 2.46 Comparison of gas tungsten and plasma arc welding processes. (*Source:* Reproduced by permission of American Welding Society, ©*Welding Handbook*).

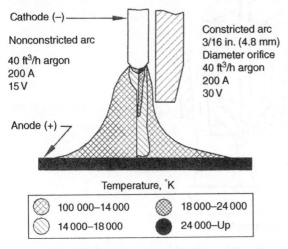

Cathode (−)

Nonconstricted arc

40 ft³/h argon
200 A
15 V

Constricted arc
3/16 in. (4.8 mm)
Diameter orifice
40 ft³/h argon
200 A
30 V

Anode (+)

Temperature, °K

100 000–14 000		18 000–24 000
14 000–18 000		24 000–Up

Figure 2.47 Effect of arc constriction. (*Source:* Reproduced by permission of American Welding Society, ©*Welding Handbook*).

accomplish manually. This fact combined with the relatively bulky torch are reasons why most of the applications for this process are mechanized.

There are two PAW modes—transferred and nontransferred (Figure 2.49). With the transferred mode, the arc is formed between the electrode and the work. This results in the greatest arc plasma energy density and weld penetration, and therefore, is by far the most common mode used in the industry. In nontransferred PAW, the arc forms between the electrode and the base of the constricting nozzle (which has a wider opening), resulting in an arc that has much lower energy density.

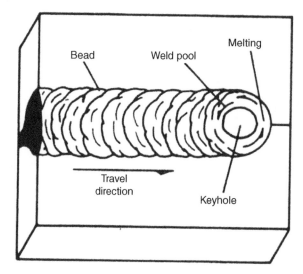

Figure 2.48 Keyhole mode welding. (*Source:* Reproduced by permission of American Welding Society, ©*Welding Handbook*).

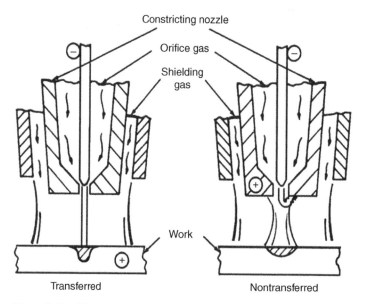

Figure 2.49 The two modes of plasma arc welding. (*Source:* Reproduced by permission of American Welding Society, ©*Welding Handbook*).

This method is useful for welding very thin work pieces when much lower heating is needed. It can also be used for cutting of nonconductive materials.

DCEN (pulsed or nonpulsed) is used most often with PAW, but square wave AC may be used for aluminum and magnesium to take advantage of the cleaning action from the positive half of the cycle. Typical current ranges are similar to GTAW—less than 1 ampup to 500amps. Due to much longer arc lengths, it is not surprising that PAW arc voltages tend to be much higher than GTAW arc voltages, often exceeding 30 volts. PAW arcs are initiated by a pilot arc established between the electrode and the copper alloy constricting nozzle. Pilot arcs typically use high frequency currents that are generated by an additional power source integrated into the system.

Compared to the GTAW process, PAW offers the following advantages and limitations:

Advantages:

- Higher energy density plasma column provides for greater penetration, faster welding speeds, and less overall heat into the part
- Due to the columnar shaped arc, heating from the arc is less affected by arc length variations
- Electrode contamination is reduced so the electrode wear is much less

Limitations:

- Equipment is more expensive
- Keyhole mode welding is difficult to perform manually
- Torch is bulky

2.6 Gas Metal Arc Welding

Gas Metal Arc Welding (GMAW) uses a continuously fed bare wire electrode through a nozzle that delivers a blanketing flow of shielding gas to protect the molten and surrounding hot base metal as it cools (Figure 2.50). It is commonly referred to by its slang name "MIG" (Metal Inert Gas) Welding. Because the wire is fed automatically by a wire feed system, GMAW is considered to be a semiautomatic process. The wire feeder pushes the electrode through the welding torch where it makes electrical contact with a copper contact tube which delivers the current from the power supply. Because arc length is controlled by the power supply, the process requires less welding skill than SMAW or GTAW, and produces higher deposition rates. Like GTAW, it is sensitive to welding in drafty conditions.

Figure 2.51 shows a typical GMAW machine setup. The basic equipment components are the welding gun and cable assembly, electrode feed unit, power supply, and source of shielding gas.

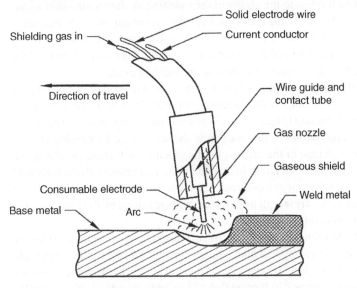

Figure 2.50 Gas metal arc welding. (*Source: Welding Essentials*, Second Edition).

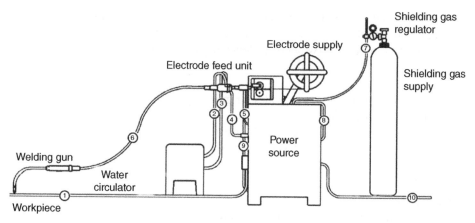

Figure 2.51 Typical equipment set up for gas metal arc welding. (*Source:* Reproduced by permission of American Welding Society, ©*Welding Handbook*).

This setup shows a water-cooling system for the welding gun, which is typically used when welding with high duty cycles and high current.

GMAW became commercially available in the late 1940s, and offered a significant improvement in deposition rates, making welding a more efficient fabrication process than ever before. GMAW is readily adaptable to robotic applications. Since the wire feed is continuous, overall welding times are greatly reduced as compared to the manual processes since there is no need to constantly stop and start to replace filler metal. Because of the fast welding speeds, high deposition rates, and the ability to adapt to automation, it is widely used by automotive and heavy equipment manufacturers, as well as by a wide variety of construction and structural welding, pipe and pressure vessel welding, and cladding applications. It is extremely flexible and can be used to weld virtually all metals. Relative to SMAW, GMAW equipment is a bit more expensive due to the additional wire feed mechanism, more complex torch, and the need for shielding gas, but it is still considered a relatively inexpensive welding process.

GMAW is "self-regulating," which refers to the ability of the machine to always maintain a constant arc length. This is usually achieved using a constant voltage power supply, although some modern machines are now capable of achieving self-regulation in other ways. This self-regulation feature makes the process ideal for mechanized and robotic applications since there is no need for the operator to control arc length, as is the case with manual arc welding processes.

As mentioned previously in this chapter, self-regulation with a constant voltage power supply takes advantage of the relatively flat volt-ampere characteristic curve, and the relationship between welding current and electrode melting rate (Figure 2.52). When the arc voltage is set by the operator, the power supply monitors and automatically maintains the set voltage by keeping the arc length constant. Upon any abrupt change in the arc length, the machine will sense the change in voltage and react by rapidly changing the current as indicated on the volt-ampere characteristic of the figure. As explained before, the subsequent rapid changes in current produce rapid changes in electrode melting rate, which adjust the arc length until the arc voltage sensed by the power supply is equivalent to the machine set point voltage.

Figure 2.53 provides important GMAW terminology. Of particular importance is electrode extension. As shown, electrode extension refers to the length of filler wire between the arc and the end of the contact tip. The reason for the importance of electrode extension is that the longer the electrode extension, the greater the amount of resistive (known as I^2R) heating that will occur in the wire. Resistive heating occurs because the steel wire (which delivers the current to the arc from the contact tip) is a relatively poor conductor of electricity. This effect can become significant at high currents and/or long extensions, and can result in more of the energy from the power supply being consumed in the heating of the wire, and less in generating arc heating. As a result, significant resistive heating can result in a wider weld profile with less penetration or depth of fusion. The standoff distance is also an important consideration. Distances that are excessive will adversely affect the ability of the shielding gas to protect the weld, while distances that are too close may result in excessive spatter buildup on the nozzle and contact tip.

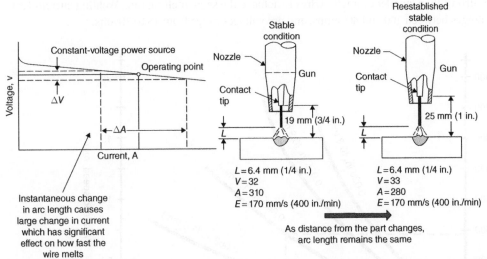

Figure 2.52 Gas metal arc welding self-regulation. (*Source:* Reproduced by permission of American Welding Society, ©*Welding Handbook*).

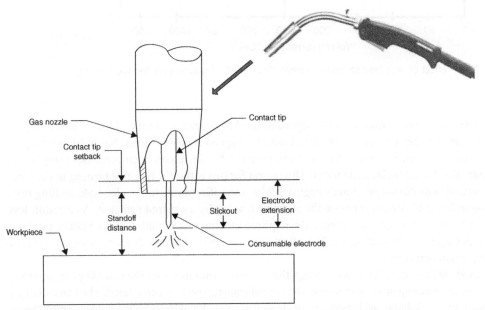

Figure 2.53 Common gas metal arc welding terminology. (*Source:* Reproduced by permission of American Welding Society, ©*Welding Handbook*).

In addition to the welding variables common with all arc welding processes such as voltage, current, and travel speed, a semiautomatic process such as GMAW includes the additional important variable of wire feed speed. Wire feed speeds affect both deposition rates and current. As described previously, as wire feed speeds increase, the machine automatically must increase the current to melt the wire faster to maintain the same arc voltage (or arc length). As a result, current and wire feed speed are interrelated, so the current adjustment and wire feed adjustment are effectively the same. The amount of current needed per given wire feed speed will be a function of the diameter of the wire, as indicated in Figure 2.54. At a given wire feed speed, larger diameter wires require more current than smaller diameter wires to achieve the same melting rate. Welding current typically ranges between 100 and 400amps, and arc voltages range from 15 to 30volts.

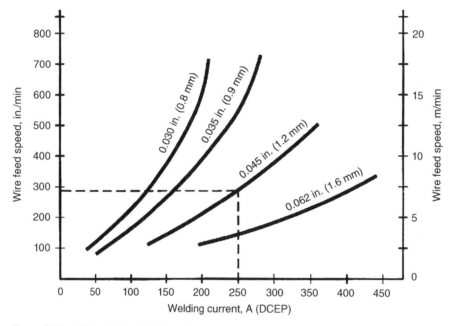

Figure 2.54 Effect of wire feed speed on current. (*Source: Welding Essentials*, Second Edition).

An interesting observation from this figure is the slight curvature to the plots. At lower amperages, the relationship between current and wire feed speed is nearly linear. But at higher amperages, increases in wire feed speed begin to require smaller increases in current, especially when using smaller diameter electrode wires. The reason for this is the I^2R resistive heating in the electrode extension discussed previously begins to play a significant role in the electrode melting rate. Increasing levels of current increase the resistive heating by a squared function. As a result, less additional current is needed to achieve the required melting rate to maintain arc voltage (and arc length) per a given wire feed speed. The effect is greater in the smaller diameter wires because of higher current densities.

The mode of filler metal transfer through the arc is very important for the GMAW process which is the only arc welding process in which filler metal transfer mode is considered. The three distinct modes are spray, globular, and short circuit, as well as a version of spray called pulsed spray (Figure 2.55). The type of metal transfer mode influences many aspects of the process. Those include weld metal deposition rate and travel speed, heat input to the part, welding positions possible, part thicknesses that can be welded, and the ability to weld in gaps.

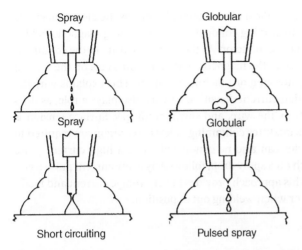

Spray   Globular

Spray   Globular

Short circuiting   Pulsed spray

Figure 2.55 Metal transfer modes with gas metal arc welding.

Spray transfer mode is the common choice for high-production welding. It produces significant depth of fusion and high deposition rates. This mode is characterized by a fine dispersion of tiny drops that do not create short circuits in the arc, resulting in less spatter than other modes. For a given wire diameter, a transition from globular transfer mode to spray transfer mode will occur quite abruptly at a specific current (Figure 2.56). The current at which this transition occurs is known as the spray transition current. Current levels above the transition current will produce spray transfer, while current levels below the transition current produce globular transfer. Larger electrode diameters require higher transition currents. As the figure indicates, the transition to spray transfer is associated with a substantial increase in the rate of drop transfer. Obviously, this transition is also associated with a significant decrease in the volume of each drop (relative to globular transfer). Spray transfer mode will only occur when the shielding gas consists of at least 80% argon. It is mainly limited to flat and horizontal welding positions, although pulsing provides greater heat and puddle control, allowing for the possibility to weld out-of-position.

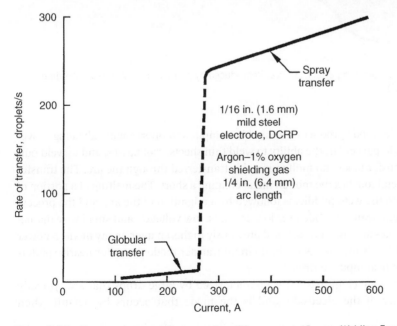

Figure 2.56 Gas metal arc welding spray transition current. (*Source: Welding Essentials*, Second Edition).

The spray transfer transition is directly related to a phenomenon known as the electromagnetic pinching effect (Figure 2.57). In any current-carrying conductor, a surrounding magnetic field is created which produces an inward force on the conductor. As the current density in the conductor increases, so does the inward squeezing force of the magnetic field. This force acts on the molten drops forming on the end of the wire by squeezing or pinching them off. This explains why the spray transfer mode requires sufficiently high current densities. Gravity also plays a role, as does the surface tension between the molten drop, the wire, and surrounding gas. Surface tension is affected by the type of gas which is why gas mixtures containing at least 80% argon are required to achieve spray transfer. Pulsed spray transfer can also be conducted using a high current pulse (above the spray transition current) to pinch off a spray drop, followed by a low current pulse (typically below the spray transition current). This approach provides lower average current and facilitates using spray transfer on thin material or when welding out of position.

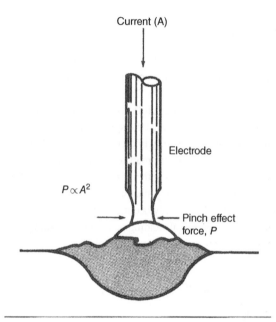

Current (A)

Electrode

$P \propto A^2$

Pinch effect force, P

Figure 2.57 Electromagnetic pinching effect. (*Source:* Reproduced by permission of American Welding Society, ©*Welding Handbook*).

The short circuit transfer mode produces minimal heat input and offers many advantages over the spray transfer mode, in particular, the ability to weld thin sheets, root passes, and to weld out-of-position. With this transfer mode, no molten metal is transferred through the arc. The transfer occurs when the filler metal touches the molten metal creating a short. The melting of a portion of the wire, which flows into the weld puddle, is followed by a reignition of the arc, and the process continues. The short circuit transfer mode uses low currents, low voltages, and small wire diameters, and the deposition rates are low. As discussed previously in this chapter, many modern power supplies offer considerable advancements in short circuit transfer mode that offer nearly spatter-free welding with precise heat input control.

The globular transfer mode (Figure 2.58) is characterized by large drops that are typically larger than the diameter of the electrode, and is the mode that occurs by default when conditions are not right for spray or short circuit transfer. Since this transfer mode is dominated by gravity, it is mainly limited to flat and horizontal welding positions. The globular transfer mode uses lower currents and produces less part heating than spray transfer, but it produces significant amounts of spatter and a rough weld bead appearance. As a result, it generally is not the optimum choice for production but does offer the significant advantage of working well with 100% CO_2 gas, which is much less expensive than argon. A variation of globular transfer with relatively high current and low voltage is known as buried arc. The arc force from the higher current serves to depress the weld pool, allowing the tip of the electrode to be in a cavity surrounded by the molten metal. This approach tends to contain spatter and force molten weld metal out to the edges, facilitating faster travel speeds without producing "ropey" weld beads.

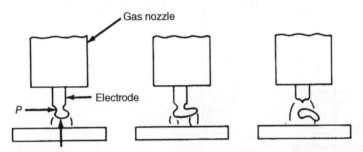

Figure 2.58 Globular transfer mode. (*Source:* Reproduced by permission of American Welding Society, ©*Welding Handbook*).

The choice of shielding gas for the GMAW process can significantly affect the weld profile due to effects on arc heating, shape, and stability, as well surface tension of the puddle and droplets. As mentioned previously, at least 80% argon is required for the spray transfer mode. The argon arc also exhibits a wide temperature range from the very hot center to the relatively cool temperatures toward the outer portions of the arc. This results in a weld profile that is relatively narrow and deep penetrating. The higher surface tension created with the argon arc produces excellent wetting with the base metal. Pure argon is typically used for nonferrous metals, while argon blends containing CO_2, helium, and/or oxygen are used for steel and stainless steel. Argon blends containing small amounts of oxygen can further affect surface tension of the weld puddle and reduce the tendency for a weld metal discontinuity known as undercut.

Helium arcs operate at higher voltage and are hotter with higher thermal conductivity than argon arcs resulting in wide, deep penetrating weld profiles good for welding thick materials. Helium arcs are rougher and produce more spatter, and typically require some argon additions to improve arc stability. When helium is used while welding in the flat position, the flow rate will generally need to be two to three times greater than argon since helium is much less dense than air. Helium and helium blends are typically used for nonferrous metals.

Although it is not inert, CO_2 is often used when welding carbon steel. It also produces a rough arc with significant spatter and a weld bead that is not always esthetically pleasing, but with excellent depth of fusion. As mentioned before, the main advantage of this gas is its low cost (it is the same gas used in carbonated beverages). Mechanical properties, however, may not be as good as welds made with inert gas because of oxidation from the CO_2 gas. In summary, a wide variety of gasses and gas blends are used with GMAW depending on the application, the desired metal transfer mode, and the metals being welded.

Welding technique and the torch travel angle can significantly affect the bead profile as well. As indicated in Figure 2.59, the backhand technique will generally produce deeper penetration and a narrower weld profile with greater convexity than the forehand technique. A drag angle (with the backhand technique) of about 25° will produce the greatest penetration, but angles of 5–15° will generally provide the best control of the puddle and the most optimum shielding. Ultimately, the choice of torch angle depends on many factors, including material type and thickness being welded, welding position, desired weld profile, and welder preference.

AWS A5.18 is the carbon steel filler metal specification for Gas Shielded Arc Welding and includes filler metal for both GMAW and GTAW. A typical electrode specification is shown in Figure 2.60. The E refers to electrode and the R refers to rod, which means that the filler metal can

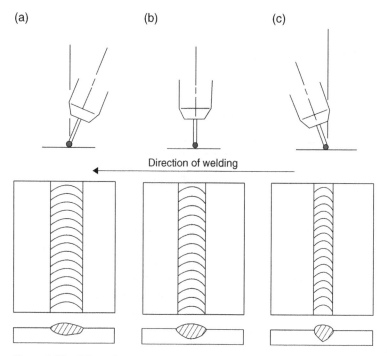

Figure 2.59 Effect of welding technique on weld profile. (a) Forehand technique, (b) torch perpendicular, and (c) backhand technique. (*Source:* Reproduced by permission of American Welding Society, ©*Welding Handbook*).

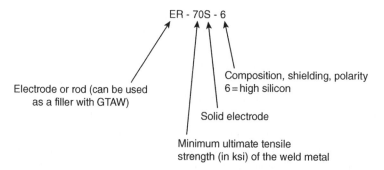

Figure 2.60 Typical AWS A5.18 filler metal specification.

be used either as a GMAW electrode (which carries the current), or as a separate filler metal (rod) that can be used for the GTAW process. The "S" distinguishes this filler metal as solid, whereas a "C" means it is a composite electrode. Composite electrodes are also known as metal cored wires, and are becoming more popular with the recent advancements in filler metal development. They offer high deposition rates without the excessive fumes and slag associated with FCAW. The number, letter, or number/letter combination, which follow the "S" or "C" refers to a variety of information about the filler metal such as composition, recommended shielding gas, and/or polarity the filler metal is designed for.

In summary, the GMAW process offers the following advantages and limitations:

Advantages:

- Higher deposition rates than SMAW and GTAW
- Better production efficiency versus SMAW and GTAW since the electrode or filler wire does not need to be continuously replaced
- Since no flux is used there is minimal postweld cleaning required and no possibility for slag inclusions
- Requires less welder skill than manual processes
- Easily automated
- Can weld most commercial alloys
- Deep penetration with spray transfer mode
- Depending on the metal transfer mode, all position welding is possible

Limitations:

- Equipment is more expensive and less portable than SMAW equipment
- Torch is heavy and bulky so joint access might be a problem
- Various metal transfer modes add complexity and limitations
- Susceptible to drafty conditions

2.7 Flux Cored Arc Welding

Flux Cored Arc Welding (FCAW) is much like GMAW and uses similar equipment but relies on a tubular filler wire containing a variety of materials such as powdered metal and alloying elements, materials that decompose to form a gas and melt to form a slag, deoxidizers, and scavengers (Figure 2.61). There is no need for binding agents to hold the flux material to the wire as is the case with SMAW electrodes. As a result, there is more space to add elements to the flux that produce optimum welding conditions such as high deposition rates, excellent welding characteristics, and improved weld metal properties.

Because of the flexibility it offers and high deposition rates, in addition to continuous improvements in available filler materials, FCAW is a popular choice as a replacement for GMAW and/or SMAW for many applications. It offers the possibility for welding with or without shielding gas. Equipment costs are similar to GMAW, but the filler wire is more expensive, and fume generation can be very high, especially in the self-shielded version of FCAW.

There are two process variations associated with FCAW—a self-shielded version (Figure 2.61) and a gas-shielded version (Figure 2.62). The self-shielded version contains material in the flux that decomposes to form a shielding gas. Since this gas forms immediately in and around the arc, it is less sensitive to being displaced by drafty conditions than the gas-shielded version, and

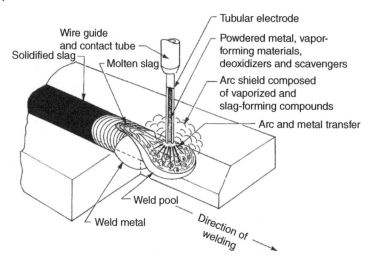

Figure 2.61 Flux cored arc welding. (*Source: Welding Essentials*, Second Edition).

therefore, is more tolerant to welding in the field. And since pressurized gas cylinders are not needed, the self-shielded version is also more portable. The gas-shielded version uses CO_2 gas or a blend of CO_2 and argon gas for shielding. The advantage of the gas-shielded version is less fume production and a greater ability to add beneficial materials to the flux since there is no need for adding additional material to form a shielding gas. The mechanical properties of gas-shielded welds produced with this process may be better in many cases as well.

Figure 2.63 shows a simple version of the FCAW electrode classification system. As compared to other electrodes, only one number is used to designate the minimum tensile strength of the deposited weld metal, so its units are in 10,000 pounds per square inch. The second number refers to the

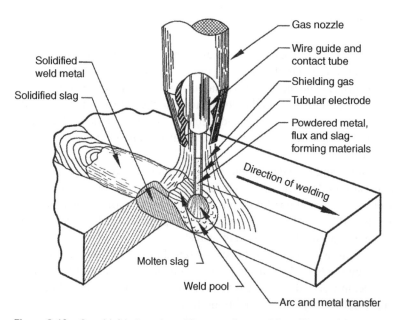

Figure 2.62 Gas-shielded version of flux cored arc welding. (*Source: Welding Essentials*, Second Edition).

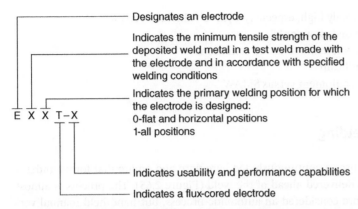

Figure 2.63 Flux cored arc welding classification system. (*Source:* Reproduced by permission of American Welding Society, ©*Welding Handbook*).

position(s) for which the electrode can be used, and there are only two options: 0 (flat and horizontal) and 1 (all). The "T" that follows these numbers refers to a tubular wire, clearly identifying it as a FCAW wire. Following the "T" are numbers and/or letters that provide a wide range of information including its general usability characteristics, and whether or not the electrode requires gas shielding, can be used for multi-pass welding, and meets the requirements of the diffusible hydrogen test as well as the requirements for improved toughness (the figure only shows a single "X" at the end; their actually may be many more letters than this).

The AWS carbon steel specification for FCAW electrodes is A5.20. A typical electrode is shown in Figure 2.64. Since there is no "R" following the "E," the electrode cannot be used as a rod for GTAW. Position "0" means that this electrode can only be used in flat and horizontal positions. Finally, a number "1" at the end indicates that this electrode is commonly shielded with CO_2 (but other gas blends may be substituted), can be used for both single and multiple pass welding, and uses the DCEP polarity.

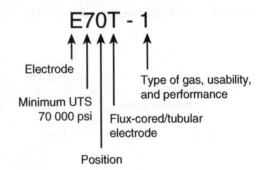

Figure 2.64 Typical AWS A5.20 electrode.

In summary, the FCAW process offers the following advantages and limitations:

Advantages (these are in addition to the GMAW Advantages):

- High deposition rates (higher than SMAW and GMAW, but not as high as SAW)
- Reduced groove angles versus SMAW means reduced volume of metal required to fill the joint
- Self-shielded version is tolerant to drafty conditions
- More tolerant to weld metal contamination than GMAW
- Slag covering can help with out-of-position welds

Limitations:

- Filler metal is more expensive than GMAW filler metal
- Requirement to remove slag after welding

- Fume production can be extremely high, especially with self-shielded version
- Limited to steels and nickel-based alloys only
- Gas-shielded version is not very tolerant to drafty conditions
- Spatter can sometimes be significant
- More complex and expensive equipment versus SMAW

2.8   Submerged Arc Welding

SAW is a welding process that uses a continuously fed bare wire and an arc that forms under a blanket of granular flux that is delivered ahead of the weld (Figure 2.65). The process is almost always mechanized and therefore considered an automatic process, but hand-held manual versions are also available. More than one consumable wire may be used (Figure 2.66). This process relies on a portion of the granular flux melting due to the heat of the arc to form a molten slag. Since it is less dense than metal, the molten slag remains at the top of the solidifying weld metal, protecting it from the atmosphere as well as assisting in its shape. Arc radiation, fumes, and spatter are contained under the flux.

The primary advantage of SAW is the extremely high weld metal deposition rates that are possible, especially when multiple wires are used. This is due to the excellent protection from the atmosphere and puddle control provided by the molten slag. As a result, SAW is very common for welding very thick plates or pipes where there is an opportunity to significantly reduce the number of passes that would be required with other arc welding processes. However, the large weld pool and amount of molten slag involved limits this process to nearly flat and horizontal positions.

SAW was developed in the 1950s to increase productivity of the fabrication of thick-walled structures when welding from one side only. The process is widely used today for girder and I-beam

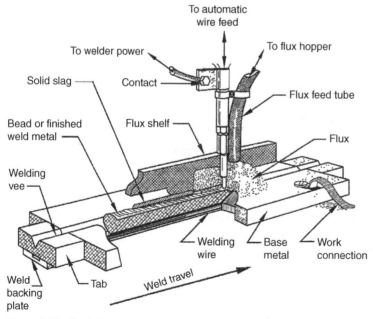

Figure 2.65 Typical mechanized arrangement of submerged arc welding. (*Source: Welding Essentials*, Second Edition).

Figure 2.66 Multiple wire submerged arc welding. (*Source:* Reproduced by permission of American Welding Society, ©*Welding Handbook*).

construction, shipbuilding, piping (Figure 2.67) and pressure vessel fabrication, rail car construction, and cladding (Figure 2.68). It is easily adapted to automation, but generally not used in the field due to the need for flux hoppers and collection systems. In many cases, the welding torch is stationary, requiring that the parts to be welded are moved below the torch.

The process normally operates at very high current levels (up to 2000 ampsor more) and high-duty cycles. Thus, the power supplies required for SAW are more expensive than those for competing processes such as GMAW. Arc voltages are higher than other arc welding processes, and usually range between 25 and 40 volts. Process variables are the same as semiautomatic processes such as GMAW, with the additional variable of the granular flux. Because of the very high levels of current possible, resistive heating associated with electrode extension can be even more pronounced with SAW than it is with processes such as GMAW. Table 2.7 shows typical ranges of current used as a function of the electrode diameter. As the table indicates, SAW electrode diameters can be quite large.

SAW positions for welding pipe sections together are shown in Figure 2.69. When welding on the outside of the pipe, a position slightly below the uphill (relative to the pipe rotation) side produces the best results. This allows the weld and molten slag to solidify as both materials reach the horizontal position at the top of the pipe, producing the best weld shape. If there is insufficient displacement on the uphill side, the molten weld metal and slag will spill forward, and weld solidification will occur on the downhill side of the pipe resulting in a narrow weld with excessive reinforcement. Excessive displacement will cause the flux and molten slag to spill ahead of the weld resulting in the shape shown. Similar trends when welding on the inside of the pipe are also shown. When welding plates in the flat position, it is possible to produce welds on plates that are not perfectly flat. However, as is the case when welding pipes, if the weld and molten slag do not solidify in a flat position, the weld shape will be affected (Figure 2.70).

For SAW, AWS specifies both the flux and the filler metal in a single specification. The carbon steel specification for electrodes and fluxes is AWS A5.17. For example, a common specification for a flux–electrode combination from this specification is F7A2-EM12K. The "F" indicates that this is a SAW flux. The information after the "F" refers to the weld metal properties that a weld made with this flux–electrode combination will exhibit, and what heat treatment (if any) is required to produce these properties. The first number is the minimum deposited metal tensile

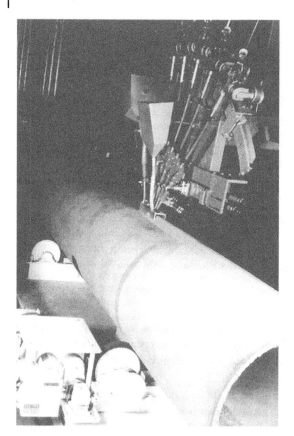

Figure 2.67 Submerged arc welding the seam of a pipe. (*Source:* Reproduced by permission of American Welding Society, ©*Welding Handbook*).

Figure 2.68 Dual head submerged arc cladding operation. (*Source:* Reproduced by permission of American Welding Society, ©*Welding Handbook*).

Table 2.7 Current ranges as a function of electrode diameter.

Wire diameter		Current range
Millimeters	Inches	Amperes
1.6	1/16	150–350
2.0	5/64	200–500
2.4	3/32	300–600
3.2	1/8	350–800
4.0	5/32	400–900
4.8	3/16	500–1200
5.6	7/32	600–1300
6.4	1/4	700–1600

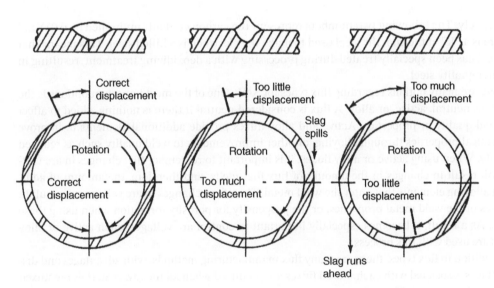

Figure 2.69 Submerged arc welding positions for pipe. (*Source:* Reproduced by permission of American Welding Society, ©*Welding Handbook*).

strength in increments of 10,000 pounds per square inch. The letter following this number will be either an "A" or a "P" which provides information regarding the heat treatment condition of the weld test plate prior to testing. An "A" indicates no heat treatment, or as-welded, and a "P" indicates that a postweld heat treatment was conducted. The number following the heat treatment designation provides information regarding Charpy Impact properties of the weldment when produced with the electrode shown. Specifically, the temperature (in °F multiplied by −10) at which a minimum of 20 ft-lb is achieved via Charpy Impact testing. For example, if the number is a 2, the Charpy Impact testing temperature is −20°F, if it is a 4 the temperature is −40°F, and so on. The designations for the electrode refer to the chemistry of the electrode that is required to produce the weld metal properties described under the flux designation. A "C" refers to a composite electrode while an absence of a "C" means that a solid wire is being used. The next letter, an "L," "M," or "H" refers to low, medium, or high manganese level,

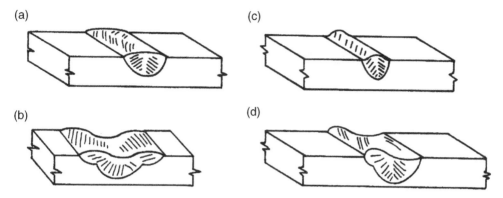

Figure 2.70 Effect of slope of plate on shape of submerged arc weld. (a) Flat position weld, (b) downhill weld (1/8 slope), (c) uphill weld (1/8 slope), and (d) lateral weld (1/19 slope). (*Source:* Reproduced by permission of American Welding Society, ©*Welding Handbook*).

respectively. The following two numbers represent the carbon content (multiplied by 0.01) and if there is a "K" at the end, the steel used to make the electrode is a killed steel. This means that the steel has been specially treated during processing with a deoxidizing treatment, resulting in a higher quality steel.

There are many options regarding flux types for SAW. One of the most important options is the flux type: neutral, active, or alloy. A flux is considered neutral if there is nothing added to affect the weld quality or properties. Active and alloy fluxes provide additional elements to improve mechanical properties through alloying, or other improvements to weld quality such as reduced porosity. When using active or alloy fluxes, it is important to remember that changes in arc voltage will result in changes in the amount of flux that melts, and therefore, the amount of additional alloying elements that enter the weld metal. Therefore, changes in arc voltage can result in changes in the weld metal properties, or the propensity for porosity formation when using these fluxes. As a result, it becomes especially important to control arc voltages when active or alloy fluxes are used with this process.

In addition to flux types, there are many flux manufacturing methods, with advantages and disadvantages associated with each. Fused fluxes are produced when all the raw materials are mixed first and then melted. These fluxes are moisture resistant and easy to recycle but are limited in chemistry and capabilities because of the requirement for alloying elements to have similar melting temperatures. Agglomerated or bonded fluxes are dry mixed, bonded, and baked. A wide variety of alloying elements and deoxidizers can be added to these fluxes, but they are more sensitive to moisture pick-up, and may also suffer from lack of chemical homogeneity. Mechanically mixed fluxes are a mixture of both bonded and fused fluxes and offer a wide variety of weld metal properties. However, they are sensitive to segregation during shipping. Fluxes that are derived from weld slag that are removed and collected to be crushed and used again are known as crushed slag. But these fluxes may not be permitted by the applicable code or welding specification.

In summary, the SAW process offers the following advantages and limitations:

Advantages:

- Extremely high deposition rates
- No arc radiation
- Minimal smoke and fumes

- Significant opportunity to customize weld metal properties through the selection of the flux
- Welds can be produced with reduced groove angles
- Mechanized process (usually) does not depend on welder skill

Limitations:

- Restricted to flat position for groove welds, and flat and horizontal positions for fillet welds
- Flux handling equipment adds complexity
- Requirement to remove slag
- Flux can easily absorb moisture which increases hydrogen cracking susceptibility when welding certain steels
- Not suitable for thin sections

2.9 Other Arc Welding Processes

2.9.1 Electrogas Welding

Electrogas Welding (EGW), a variation of Electroslag Welding (described next), is a mechanized arc welding process that is designed for welding vertical seams in a butt joint configuration between extremely thick sections (Figure 2.71). It uses a continuously fed wire that may be solid or flux cored. Shielding gas (typically CO_2) is required when solid wire is used. The cored wire version is usually much like the self-shielded version of FCAW where no additional gas is required. The molten material is typically contained in the joint by a set of water-cooled copper dams, although ceramic versions are sometimes used. The process is similar to a casting process in that it involves filling the joint with a pool of weld metal beginning at the bottom of the joint and progressing to the top. A starting tab is required at the bottom of the joint and run-off tabs may be used at the top. Although the overall movement of the weld pool is vertical up, the actual weld position is flat. Deposition rates are the highest of all arc welding processes, and heat input is extremely high.

There are two versions of the process—one in which the water-cooled copper dams and the welding apparatus move vertically with the rising weld, the other in which the dams remain stationary. The stationary method uses a consumable guide tube that melts with the electrode as the pool continues to rise in the joint. The consumable guide tubes match the base metal chemistry, and account for as much as 30% of the deposited weld metal. The stationary method is mainly applicable to joints of relatively small length (<5 ft).

A common application for EGW is the welding of thick panels to form the hulls in shipbuilding (Figure 2.72). As the figure indicates, with large ships the seams can be very long. Large storage tanks and vessels are another common application. The main advantage of Electrogas over Electroslag Welding is reduced heat input, and therefore, reduced grain sizes and improved mechanical properties (especially toughness) of the joint.

2.9.2 Electroslag Welding

Electroslag Welding (Figure 2.73) is similar to EGW, except that Electroslag Welding relies completely on a molten slag, so there is no arc once the process is initiated. The heat that causes filler and base metal melting comes from resistive heating of the molten slag as welding current passes through it. A granular flux is used much like the flux associated with the SAW process. An arc forms at the beginning of the process, melting a portion of the flux to form the molten slag, which protects the

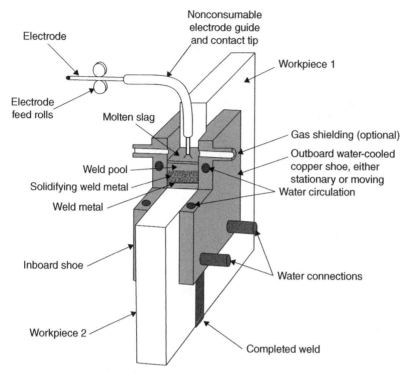

Figure 2.71 Electrogas welding. (*Source:* Reproduced by permission of American Welding Society, ©*Welding Handbook*).

Figure 2.72 Electrogas welding of ship hulls. (*Source:* Reproduced by permission of American Welding Society, ©*Welding Handbook*).

solidifying weld. The arc is then quickly extinguished by the molten slag puddle, and the weld continues to progress via the heat supplied by the resistive heating of the slag pool. As a result, Electroslag Welding is truly not an arc welding process even though it is often categorized as one.

Electroslag Welding was first used in the United States in 1959. Potential applications, much like EGW, include any fabrication requiring the welding (splicing) of long vertical seams between thick-walled plates. It is also used for cladding.

In summary, the Electrogas and Electroslag processes offer the following advantages and limitations:

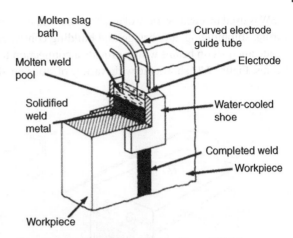

Figure 2.73 Electroslag welding. (*Source: Welding Essentials*, Second Edition).

Advantages:

- Extremely high deposition rates
- Ability to rapidly weld the vertical seams of very thick plates in a single pass
- No interpass cleaning required since weld is completed in a continuous operation

Limitations:

- Mainly limited to carbon and low-alloy steels, and some stainless steels
- Requires vertical butt joint design and straight seams
- Not applicable to thin plates
- Extremely high heat inputs (especially with Electroslag) create large grains and reduced mechanical properties such as toughness
- Once the weld sequence has started, it must continue until completion or weld defects may be created

2.9.3 Arc Stud Welding

Arc Stud Welding (SW) is an arc welding process designed to attach a stud quickly and efficiently to a plate. The welded stud typically acts as a threaded attachment to provide for connections to other parts or an anchor to connect steel beams to concrete. The process (Figure 2.74) uses a constant current power supply, and a gun that positions the stud for welding. A ceramic ferrule protects the weld area from the atmosphere and helps shape the weld.

The weld sequence is shown in Figure 2.75. When the operator pulls the trigger, the stud first touches the part (a) and then is drawn away (b), while voltage is applied initiating an arc. This is referred to as a drawn arc start. The heat of the arc melts the end of the stud as well as the base metal below the stud. Steel studs typically contain a small ball of aluminum that is press-fit into a hole at the end of the stud. Upon arc initiation, the ball of aluminum quickly melts and vaporizes. The vaporized aluminum acts as a deoxidizer to protect the molten metal of the stud and base metal. The stud is then rapidly plunged into the base metal (c) which creates the weld while it expels contaminants and oxidized material through the ferrule holes and squeezes out molten metal into the ferrule cavity. The entire weld sequence typically lasts less than a second. The final step involves removing the ceramic ferrule by breaking it with a hammer to reveal the weld (d). The process requires no special operator skill.

SW is used in a wide variety of applications and industry sectors, including automotive manufacturing, shipbuilding, and bridge and building construction. Figure 2.76 depicts the use of SW to create anchors on beams which provide a connection to the concrete that is later poured onto the beams. Figure 2.77 shows the wide variety of studs possible.

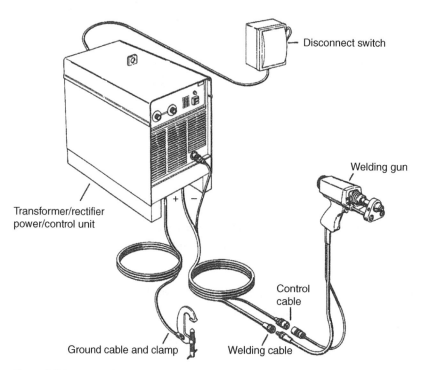

Figure 2.74 Arc stud welding set-up. (*Source:* Reproduced by permission of American Welding Society, ©*Welding Handbook*).

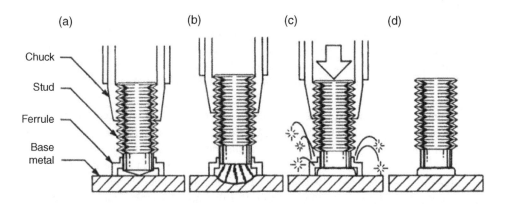

Figure 2.75 Arc stud welding steps—(a) stud contacts part, (b) arcing commences, (c) creation of weld, contaminates expelled from ferrule, (d) completed weld. (*Source:* Reproduced by permission of American Welding Society, ©*Welding Handbook*).

Figure 2.76 Arc stud welding of anchors. (*Source:* Reproduced by permission of American Welding Society, ©*Welding Handbook*).

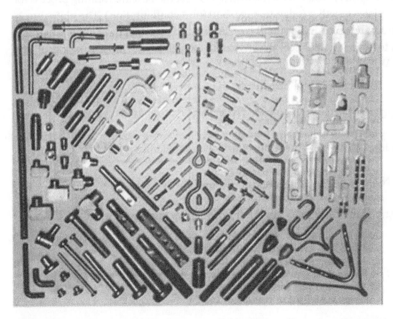

Figure 2.77 A wide variety of studs are available. (*Source:* Reproduced by permission of American Welding Society, ©*Welding Handbook*).

2.10 Test Your Knowledge

I) Fundamental Concepts—True/False

The following true/false questions pertain to some of the most important fundamental concepts in this chapter:

1) Arc voltage and open circuit voltage of the power supply refer to the same thing.
2) A welding arc can be described as an ionized gas.
3) "Self-regulation" of the arc length when using a constant voltage power supply happens because the power supply is constantly adjusting wire feed speed.
4) The GMAW process utilizes a constant current power supply.
5) When welding with GTAW, DCEN (straight polarity) delivers most of the arc heat to the part.
6) The SMAW process is known to work well with all metals, including titanium and aluminum.
7) In order to achieve spray transfer when welding with the GMAW process, the shielding gas must contain at least 80% argon.
8) When welding with the SMAW process, both DCEP and DCEN polarities can be used, depending on the electrode type.
9) The SAW process works well in all positions.
10) The letters "ER" in the AWS A5.18 filler metal specification for the ER-70S-6 filler metal refer to the fact that this wire can be used both as a filler metal rod for GTAW and an electrode for GMAW.

II) Solve a Welding Engineering Problem

The following challenge represents a typical problem Welding Engineers might encounter in their career:

In your job as a Welding Engineer, you have been asked to select the best arc welding process for welding two thick-walled plates together (edge-to-edge, butt joint configuration). The plates are 2″ thick and 15 feet long and will be welded in the flat position inside a fabrication shop. There are over 1000 of these welds to be made, so it is imperative that you choose the most economical process for this application. Which process would you choose, and why?

Recommended Reading for Further Information

ASM Handbook, Tenth Edition, Volume 6—"Welding, Brazing, and Soldering," ASM International, 1993.

AWS Welding Handbook, Ninth Edition, Volume 2—"Welding Processes, Part 1," American Welding Society, 2004.

Welding Essentials—Questions and Answers, Second Edition, International Press, 2007.

3

Resistance Welding Processes

3.1 Fundamentals and Principles of Resistance Welding Processes

Resistance welding processes represent a family of industrial welding processes that produce the heat required for welding through what is known as Joule ($J = I^2Rt$) heating. In this equation, I is current, t is time, and R is resistance. Resistance (R) of a conductor is equal to the material's resistivity (a physical property of a metal), multiplied by its length, and divided by its area. Much in the way a piece of wire will heat up when current is passed through it, a resistance weld is based on the heating that occurs due to the resistance of current passing through the parts being welded. Since steel is not a very good conductor of electricity (its resistivity is relatively high), it is easily heated by the flow of current and, therefore, is an ideal metal alloy for resistance welding processes. There are many resistance welding processes, but the most common is Resistance Spot Welding (Figure 3.1). All resistance welding processes use three primary process variables—current, time, and pressure (or force). The automotive industry makes extensive use of resistance welding, but it is also used in a variety of other industry sectors including aerospace, medical, light manufacturing, tubing, appliances, and electrical.

3.1.1 Resistance and Resistivity

The Resistance Spot Welding process is often used as a model to explain the fundamental concepts associated with most resistance welding processes. Figure 3.2 shows a standard Resistance Spot Welding arrangement in which two copper alloy electrodes apply force and pass current through the steel sheets being welded. The figure depicts the initial flow of current from one electrode to the other as an electrical circuit that contains seven "resistors." "Resistors" #1 and #7 represent the bulk resistance of the copper electrodes, #2 and #6 represent the contact resistance between the electrodes and the sheets, #3 and #5 represent the bulk resistance of the steel sheets, and #4 represents the contact resistance between the sheets. These resistances depict the typical variations in resistance across the joint when current first begins to flow and prior to weld formation. When describing these different resistances, the word "bulk" is used to make the distinction between the resistance in the material (electrodes or steel sheets) itself versus the contact resistances at the mating surfaces.

Welding Engineering: An Introduction, Second Edition. David H. Phillips.
© 2023 John Wiley & Sons, Inc. Published 2023 by John Wiley & Sons, Inc.
Companion Website: www.wiley.com/go/Phillips/WeldingEngineeringIntroduction

Figure 3.1 Resistance spot welding.

An important characteristic that is critical to most resistance welding processes is the contact resistance between the parts (or sheets) being welded. As indicated in Figure 3.2, the highest resistance to the flow of current occurs where the sheets come into contact with each other. This high resistance promotes the formation of a weld (known as a nugget) exactly where it is needed—between the parts.

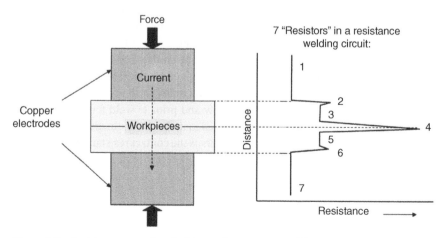

Figure 3.2 Resistances associated with a resistance spot weld.

Figure 3.3 Resistivity of steel rises rapidly with increasing temperature. (Adapted from *Resistance Welding Fundamentals and Applications* (CRC), figure 2.2, page 21).

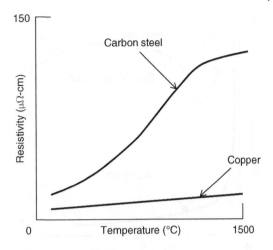

Another characteristic that works strongly in favor of resistance welding of steel is that as the steel is heated, its resistivity relative to copper increases rapidly (Figure 3.3). With most resistance welding processes, initial contact resistance begins the weld process by rapidly heating the region where the sheets parts touch. Further rapid heating to form the nugget is helped by the fact that the resistivity of the steel is now higher due to its increased temperature from the contact resistance heating. So, essentially, the contact resistance initiates the increase in the bulk resistance of the steel, which in turn provides for further heating promoting rapid growth of the weld nugget. This transition from contact resistance to bulk resistance plays an important role in many resistance welding processes. All of this occurs very fast; a typical weld time for Resistance Spot Welding of sheet steel is about 1/5th of a second. Amperages used in resistance welding are much higher than arc welding and are in the neighborhood of 10,000 ampsfor a typical spot weld on sheet steel. Other resistance welding processes may exceed 100,000 amps.

It is possible to monitor resistance during a single Resistance Spot Welding cycle and then plot the resistance as a function of time. This plot is known as dynamic resistance (Figure 3.4) since it reveals how much resistance changes over the course of a single weld. The dynamic resistance plot aids in the understanding of resistance welding fundamentals, as well as for providing a method for weld quality control. The figure shows characteristic curves when welding both aluminum and steel. On the steel curve, resistance starts out relatively high, drops, and then begins to rise again. This portion of the curve reveals the transition from contact resistance to bulk resistance discussed previously. Initially, contact resistance is high as surface oxides and rough surfaces create greater resistance to the flow of current. With enough pressure and heat, the oxides are shattered and the surface asperity peaks collapse, effectively lowering the contact resistance. As discussed, per Figure 3.3, the bulk resistance of the sheets begins to dominate following the initial contact resistance heating. The bulk resistance soon becomes great enough to generate sufficient heat to form a molten nugget. As the nugget grows, the current path becomes larger, reducing the current density, causing the resistance to plateau and then drop.

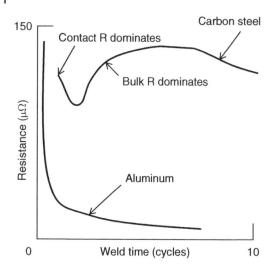

Figure 3.4 Dynamic resistance curves (one cycle = 17 milliseconds) when resistance welding steel or aluminum. (Adapted from *Resistance Welding Fundamentals and Applications* (CRC), figure 2.2, page 21).

As the figure indicates, the dynamic resistance associated with welding aluminum is completely different from that associated with steel. Initially, the resistance is extremely high due to contact resistance of the relatively thick and tenacious aluminum oxide. This contact resistance quickly diminishes as with steel, but the curve does not reveal any significant increases in bulk resistance as is the case with steel. In contrast to steel, aluminum's resistivity increases only slightly with increasing temperature, and its resistivity is only slightly higher than that of copper alloy electrodes. So, the result is bulk resistance heating does not contribute significantly to the weld formation in the way it does with steel. This is one of the many reasons why aluminum is very difficult to weld with resistance welding processes. When Resistance Spot Welding aluminum, welding currents must be much higher than for welding steel while welding times are lower, and electrode wear can be extreme.

Dynamic resistance curves can be used for monitoring quality. A curve shape that produces acceptable welds can be identified and stored in memory, and quality monitoring can then be based on pattern recognition comparisons between production curves and the stored curve. For example, if a given dynamic resistance curve does not show a significant rise in the bulk resistance portion of the curve, this might be an indication of an undersized weld. Such variations can be expected in production for a variety of reasons, including variations in the oxides and scale on the surfaces of the materials being welded.

3.1.2 Current Range and Lobe Curves

Visual examination is often an important and powerful method for verifying the quality of welds. However, with most resistance welding processes, visible examination is not possible due to the "hidden" weld location between the parts being welded. As a result, maintaining weld quality with processes such as Resistance Spot Welding are highly dependent on what is known as Current Range Curves (Figure 3.5) and Lobe Curves (Figure 3.6). They provide for a means of quality control through the monitoring of important weld parameters current and time.

The Current Range Curve is simply a plot of the spot weld diameter as a function of the welding current. It is developed experimentally by producing multiple welds using a range of current and a single time, and then measuring weld nugget diameters through a simple test known as a peel test. At current levels that are too low, unacceptable small welds will be produced. Since the

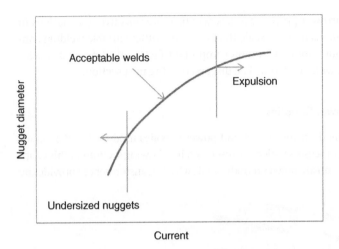

Figure 3.5 Resistance spot welding current range curve.

strength of a spot weld is highly dependent on its diameter, welds that are too small may not be strong enough for the application, and, therefore, unacceptable. Minimum weld sizes for a given application will depend on a variety of factors, including the strength of the material, the number of welds, and the loading conditions on the part. A standard rule of thumb though is that the minimum size (diameter) is equal to $(4\sqrt{t})$, where t is the sheet thickness in millimeters. At current levels that are too high, expulsion occurs. Expulsion refers to a violent ejection of molten metal from between the sheets. Although welds that incur some expulsion are often acceptable, excessive expulsion is

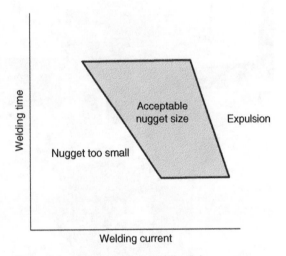

Figure 3.6 Resistance Spot Welding lobe curve.

usually undesirable because these welds tend to be very inconsistent and subject to excessive porosity and part indentation. In summary, the Current Range Curve provides a range of current that can be expected to produce acceptable welds.

The Lobe Curve incorporates welding time and represents ranges of welding current and time that will produce a spot weld nugget size that has acceptable mechanical properties for the intended application. It is produced by transposing the minimum and maximum currents from the Current Range Curves produced at various weld times resulting in what is effectively a process window for Resistance Spot Welding. Lobe Curves can be generated with different heats of incoming steel and overlapped to identify a range of parameters that work with all heats, and therefore, a process that is "robust" to heat-to-heat variations in the incoming material. During production, if a weld is made with parameters that fall outside of the Lobe Curve, the weld is considered unacceptable. Notice on the Lobe Curve shown that the window is wider at longer welding times. This is a very typical trade-off in resistance welding—production demands encourage faster welding times, but usually result in a narrower process window. A narrower process window will be more sensitive to variations in the incoming material (such as variations in oxide scale thickness) and other variable welding conditions, so quality control will be more difficult. Another important factor affecting quality with Resistance Spot Welding is electrode wear, which will be discussed in the next section.

3.1.3 Modern Equipment and Power Supplies

This discussion mainly focusses on modern equipment and power supplies used for the Resistance Spot Welding and Projection Welding processes which are discussed in following sections of this chapter. Historically, these machines have mostly relied on traditional AC/DC transformers to provide the

Figure 3.7 A modern resistance spot/projection welding machine. (*Source:* Menachem Kimchi, The Ohio State University).

current, and pneumatic cylinders to provide the pressure needed to develop electrode force. But recently, more advanced DC transformers for providing current, and servo motors for providing electrode force are becoming the dominate machines (example shown in Figure 3.7) in industry.

DC current is usually preferred because it allows for better control of the weld nugget formation. This is because with AC current, each half cycle of current passes through zero and then reaches its peak current. This cycling of current makes it more difficult to form a nugget while at the same time, minimizing expulsion. However, one problem with using DC current is due to a thermoelectric phenomenon known as the Peltier effect. As shown in Figure 3.8, which depicts a spot weld between two sheets of nickel, the

Peltier effect produces a heat imbalance toward one of the electrodes. This effect occurs when current passes through a junction of two different metals (in this case, copper and nickel). Depending on the direction of the current, more heat will be generated on one side of the spot weld than the other. In the example shown, the current travels from the top copper electrode to the top nickel sheet, and then from the bottom nickel sheet to the bottom copper electrode. These different directions from copper to nickel, and then from nickel to copper create the heat imbalance from the Peltier effect. This imbalance can result in two problems—1) a weld nugget that is not properly centered and 2)

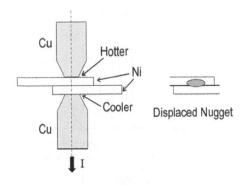

Figure 3.8 The Peltier effect can cause a heat imbalance when utilizing DC current.

excessive wear on the hotter electrode side. But one of the ways this problem is easily mitigated is by utilizing transformers that can change the DC current direction periodically to balance the heating.

The DC current associated with modern Resistance Spot Welding power supplies is generated from MFDC (medium frequency DC) power sources that are similar in design to arc welding power supplies described previously, but capable of delivering much greater levels of current. A major advantage of these power supplies for Resistance Spot Welding applications is that the transformer can be made small enough and light enough to mount on the end of a robot arm (see Figure 3.10 in the next section), and therefore, very close to the weld location. Without this capability, electrical power losses would be substantial if such large amounts of welding current (10,000 amps or greater) had to be delivered through long cables from a transformer on the factory floor to electrodes at the end of the robot arm.

Another important type of resistance welding power supply is the capacitive discharge (CD) power supply. These power supplies are more commonly used for Projection Welding applications and offer the advantage of high bursts of welding current and extremely short welding times. For Projection Welding, this offers many benefits. Welds can be made with much lower heat input which reduces distortion and marking on the side of the part without the projection. The short welding times also may provide a benefit when welding certain difficult dissimilar metal combinations by reducing the amount of time for brittle intermetallic formation.

The use of servo motors to generate the force needed for welding offers many benefits over traditional pneumatic machines. Servo motors allow for much more precise delivery and control of force through feedback control. They also make it possible to easily vary the force during a single weld. This adds an additional dimension to the process variables and can provide for better process control. Such capability may be especially helpful in difficult applications such as welding aluminum and multiple layers of dissimilar metals. Finally, there are potential benefits in the factory infrastructure to eliminating the need for pneumatically powered machines.

3.2 Resistance Spot Welding

Resistance Spot Welding (Figure 3.9) is the most common of the resistance welding processes. Its combination of extremely fast welding speeds and "self-clamping" electrodes make it an ideal process for automation (Figure 3.10) and high production environments. A typical simple AC Resistance Spot Welding weld sequence is shown in Figure 3.11. Squeeze time is the time associated with developing the proper welding force prior to passing current and is highly dependent on machine characteristics. Weld time is the time during which welding current passes. Weld time is usually very short and

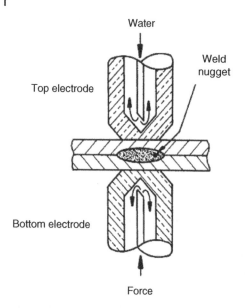

Figure 3.9 Resistance spot welding.

measured in either milliseconds or numbers of cycles if it is AC current (1 cycle = 1/60 second). Weld times vary as a function of the material being welded and the thickness, but 150–200 milliseconds is typical for steel. Although the example shows AC welding current, as mentioned previously, most modern machines now rely on DC current. Hold time is the time required to maintain electrode pressure after the current is stopped to allow for complete solidification of the weld nugget. Off time is simply the time needed to get the next part into place for welding. Quite often though the weld sequence is more complex than this simple example, and may include other features such as preheat, pulsing of DC current, and variations in the electrode force during a single weld.

Electrode geometry is a very important consideration with Resistance Spot Welding. Figure 3.12 shows some of the common shapes, but there are many more options available. In most cases, the electrodes incorporate internal channels to allow for water cooling. The electrode geometry is usually

Figure 3.10 A resistance spot welding robot. (*Source:* Reproduced by permission of American Welding Society, ©*Welding Handbook*).

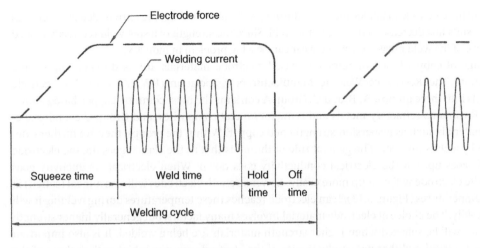

Figure 3.11 A typical resistance spot welding weld sequence. (*Source:* Reproduced by permission of American Welding Society, ©*Welding Handbook*).

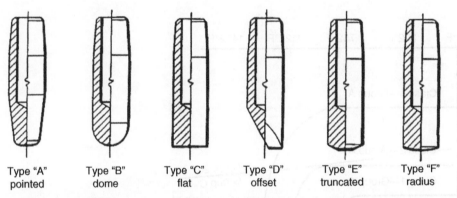

Figure 3.12 Typical resistance spot welding electrode geometries. (*Source: RWMA Resistance Welding Manual*, Revised Fourth Edition).

selected to maximize the electrical-thermal-mechanical performance of an electrode. This is generally a geometry in which the cross-sectional area increases rapidly with distance from the work piece, thereby providing a good heat sink. The choice of shape may also include considerations such as accessibility to the part and how much surface marking of the part is acceptable. For example, the type "D" offset shape is designed to place a weld close to a flange. The diameter of the electrode and the electrode contact area is also important, and is a function of the material type and thickness being welded. Too small an area will produce undersized welds with insufficient strength, while too large an area will lead to unstable and inconsistent weld growth characteristics.

Electrodes must be able to conduct current to the part, mechanically constrain the part, and conduct heat away from the part. They must be able to sustain high loads at elevated temperatures, while maintaining adequate electrical conductivity. The choice of electrode alloy for a given application is often dictated by the need to minimize electrode wear. When electrodes wear, they typically begin to "mushroom," or grow larger in diameter. Electrode wear is accelerated when there is an alloying reaction between the electrode and the part, a common problem when welding aluminum and coated steels. As the electrode diameter increases during production welding, the current density decreases which results in a decrease in the size of the weld. Since the strength of a spot weld is directly related to the size of the weld nugget, electrode wear can be a big problem in industry.

A range of copper-based or refractory-based electrode materials are used depending on the application. The Resistance Welding Manufacturers Association (RWMA) classifies electrode materials into three groups, A, B, and C. Group A contains the most common copper-based alloys, Group B contains refractory metals and refractory metal composites, and Group C contains specialty materials such as dispersion-strengthened copper. Within the groups, they are further categorized by a class number. The general rule of thumb is that as the class # goes up, the electrode strength goes up, but the electrical conductivity goes down. When electrical conductivity goes down, the electrode will heat up more easily. Lower strength electrodes will also anneal (soften) at lower temperatures (Figure 3.13). If an electrode reaches these temperatures during welding it will wear rapidly. The choice of electrode material involves many factors, but generally higher strength electrodes will be selected when higher strength materials are being welded. It is also important that the electrical and thermal conductivities of the electrode are much higher than those of the material being welded. This dictates the need for a lower-class electrode when welding highly conductive metals such as aluminum.

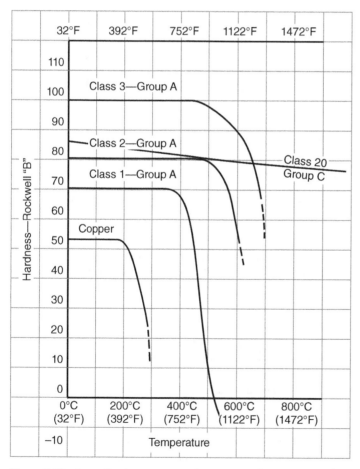

Figure 3.13 Annealing temperatures of common resistance spot welding electrode alloys. (*Source: RWMA Resistance Welding Manual*, Revised Fourth Edition).

The most used electrodes are the Class 1, 2, and 3 electrodes of Group A. Class 1 electrodes (copper with zirconium, cadmium, or chromium additions; 60 ksi UTS (cold worked); conductivity 80% IACS) offer the highest electrical and thermal conductivity and are typically used for Resistance Spot Welding of aluminum alloys, magnesium alloys, brass, and bronze. Class 2 (copper with chromium and zirconium additions, or just chromium; 65 ksi UTS (cold worked), 75% IACS) electrodes are general purpose electrodes for production Resistance Spot and Resistance Seam Welding of most materials. Class 3 (copper with cobalt, nickel, and/or beryllium additions; 95 ksi UTS (cold worked), 45% IACS) electrodes are much higher strength electrodes making them ideal for use with Projection and Flash Welding. IACS refers to a copper standard the electrodes are compared to. Pure copper has an IACS number of 100%.

As discussed, electrode wear is a big problem with Resistance Spot Welding, and ultimately results in a decrease in the size of the weld nugget (Figure 3.14). While it is not possible to completely avoid electrode wear, it is usually very important to keep it to a minimum. Factors that will increase electrode wear include excessive forces, improper electrode design or alloy type, and insufficient cooling of the electrodes. Electrode wear will be much more rapid when welding aluminum or coated steels such as galvanized steels. Approaches to addressing electrode wear include frequent tip dressing, and welding schedules that counter the effect of wear by automatically increasing the welding current after a predetermined number of welds to maintain current density.

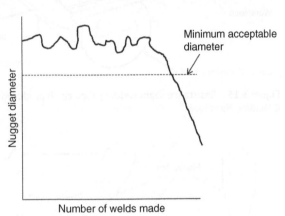

Figure 3.14 Over time, electrode wear will result in a reduction in weld size.

3.3 Resistance Seam Welding

Resistance Seam Welding (Figure 3.15) functions very similar to Resistance Spot Welding, except that the copper electrodes are rotating disks. The rotating electrodes are typically water cooled through either internal or external cooling. Much like Resistance Spot Welding, the process relies on contact resistance between the sheets, and can produce a leak-tight seam, or can be used to produce a series of spot welds. Weld speeds approaching 100 inches/min are possible. Even when leak-tight seams are being produced, the seam weld usually consists of a series of overlapping spot welds (Figure 3.16). This is achieved by pulsing the electrode current as the sheets pass between the rotating electrodes. A rule of thumb in this approach is to overlap each weld by about 30%. A continuous seam weld (as opposed to overlapping spots) is possible as well, but this approach is much more unstable and requires more robust power supplies. Like Resistance Spot Welding, Resistance Seam Welding is limited to thin sheets and joint configurations that are relatively straight or uniformly curved. Common applications include automotive fuel tanks, mufflers, and catalytic convertors.

A slightly different version of Resistance Seam Welding is known as Mash Seam Welding (Figure 3.17). This approach produces a nearly "seamless" joint and is commonly used for welding food containers (cans), water heaters, motor casings, and 30-gallon drums. Compared to conventional Resistance Seam Welding, this process relies on less joint overlap, higher forces, electrodes which are both wide and flat, and very rigid fixturing.

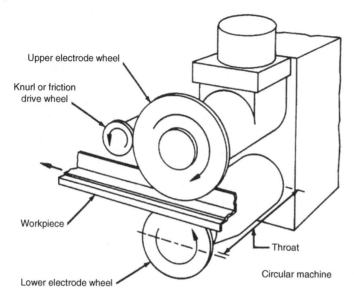

Figure 3.15 Resistance seam welding. (*Source:* Reproduced by permission of American Welding Society, ©*Welding Handbook*).

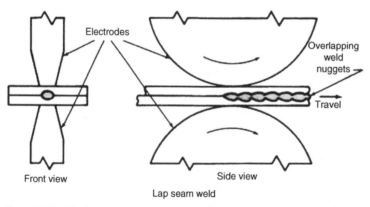

Figure 3.16 Resistance seam welds are commonly made by producing overlapping spot welds. (*Source:* Reproduced by permission of American Welding Society, ©*Welding Handbook*).

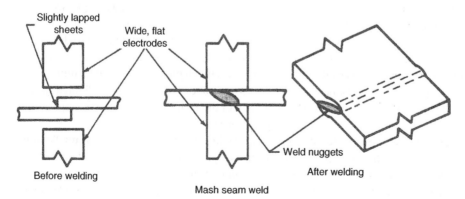

Figure 3.17 Mash seam welding produces nearly seamless joints. (*Source:* Reproduced by permission of American Welding Society, ©*Welding Handbook*).

3.4   Resistance Projection Welding

Resistance Projection Welding (Figure 3.18) is an approach to resistance welding that relies on projections machined or formed (such as stamped) on one of the parts being welded. Unique to this
process is the part projections, not the electrodes, concentrate the current/heat. A part may have only
one, two, or numerous projections. The ability for customizing the shape, number, and location of
the projections on a part creates a significant amount of design flexibility with this process. As the
figure shows, a projection weld begins with the parts first being held under pressure between a set of
copper electrodes. Current then passes through the projection(s) which softens it to collapse it. This
is followed by the final formation of the weld nugget. Some variations of the process are completely
solid-state and do not form a weld nugget.

(a)   (b)   (c)   (d)

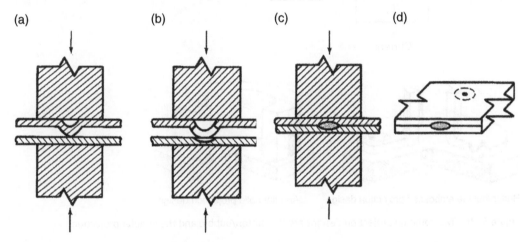

Figure 3.18   Typical resistance projection welding weld sequence beginning at step (a). (*Source:*
Reproduced by permission of American Welding Society, ©*Welding Handbook*).

As compared to Resistance Spot and Seam Welding, Resistance Projection Welding is capable of welding
much thicker parts, as well as parts with a significant thickness mismatch (Figure 3.19). As a result, it is
often considered as a potential replacement for arc welding processes such as GMAW. One of the reasons
for this is the drastic reduction in welding time that can be achieved. For example, a typical automotive part
that might require several minutes or more of welding with the GMAW process may have the potential to
be welded in less than a few seconds with the Resistance Projection Welding process. This is because the
entire weld can be made at the same time in a single fixture, and weld times are very short. Of course, this
reduction of time must be balanced against the additional expense associated with creating projections on
the part, in addition to the much greater expense of the esistance welding equipment.

There are two basic projection geometry designs (Figure 3.20): button (or bubble) and annular. Button-
type projections are typically placed at multiple locations on the part, whereas an annular projection
design may incorporate only a single projection around the periphery of the part as shown in the figure.
This approach works well for round parts. Smaller annular projections (think of small donut-shaped
projections) may also be formed and placed on multiple locations on the part.

Another distinction in projection design that is illustrated in Figure 3.20 is what is known as a solid
projection. Notice in the annular projection on the bottom of the figure that the projection is simply
an extension of the entire part and does not have a hollow cavity behind it as do the projections on
the top part. The types of projections that are extensions of the part are known as solid projections
and can only be produced by a machining or forging process, whereas the other projections are more

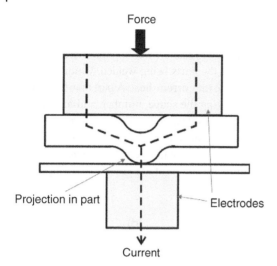

Figure 3.19 Resistance projection welding is capable of welding parts with significant differences in thickness.

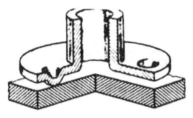

Button/bubble embossed projection design

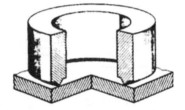

Annular solid projection design

Figure 3.20 Two common projection designs are the button/bubble and the annular projection.

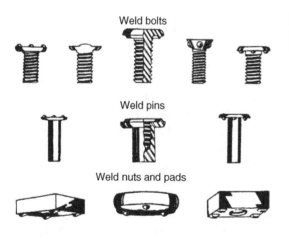

Figure 3.21 Typical projection weld fasteners. (*Source: RWMA Resistance Welding Manual*, Revised Fourth Edition).

easily produced by stamping with a punch and die. Projections produced with a punch and die usually involve the formation of a molten nugget during welding but not always. The solid projection designs always result in solid-state welds that occur via a forging action as the projection is heated and pressure applied. A very common Resistance Projection Welding application that uses solid projections involves the attachment of a wide variety of nuts, bolts, and fasteners (Figure 3.21). Many fasteners used on automobiles are attached this way.

3.5   High Frequency Welding

High Frequency Welding is a resistance welding process commonly used to weld the seams of tubes and pipes at extremely high speeds (>100 ft/min). The process uses high current at very high frequencies (200–600 kHz). There are two common versions: one is simply called High Frequency Resistance Welding (Figure 3.22) while the other is called High Frequency Induction Welding (Figure 3.23). High Frequency Resistance Welding uses two sliding electrodes in contact with the part, while High Frequency Induction Welding delivers current to the tube or pipe using an induction coil. This non-contact approach offers the advantages of eliminating electrode wear and possible marking of the tube or pipe being welded, but is not as electrically efficient as the approach using sliding electrodes.

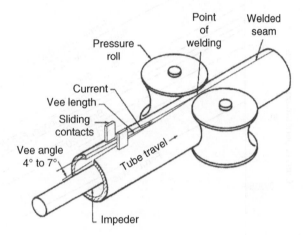

Figure 3.22   High frequency resistance welding. (*Source:* Reproduced by permission of American Welding Society, ©*Welding Handbook*).

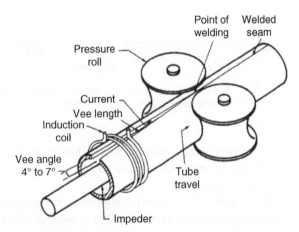

Figure 3.23   High frequency induction welding. (*Source:* Reproduced by permission of American Welding Society, ©*Welding Handbook*).

Both approaches to High Frequency Welding rely on two electrical phenomena: the "skin effect" and the "proximity effect." These two effects are associated with the extremely high frequencies of current being delivered to the weld region, and work in tandem to concentrate the current at the weld region.

The skin effect occurs when passing high frequency current through a conductor. As Figure 3.24 shows, when passing direct current (DC) through a conductor, flowing electrons would be expected to be equally distributed across the cross-sectional area of the conductor. However, when an alternating current (AC) is passed through the conductor, the electrons tend to concentrate toward the

DC   AC – 60 Hz   AC – kHz

Figure 3.24 Current path through a conductor as a function of frequency (the lighter shaded regions are void of electrons).

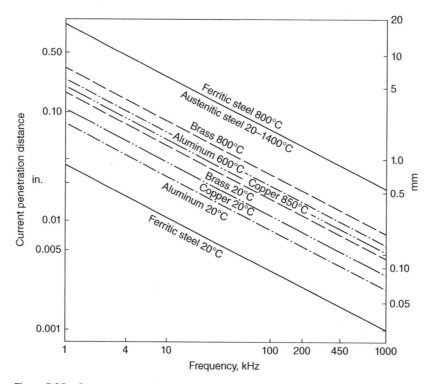

Figure 3.25 Current penetration in a conductor as a function of frequency (*Source:* Reproduced by permission of American Welding Society, ©*Welding Handbook*).

outer diameter of the conductor. The higher the frequency, the greater the effect (Figure 3.25). At very high frequencies, the current will flow only at the very surface of the conductor, hence the term "skin effect."

The proximity effect is another high frequency phenomenon, which occurs when one conductor is in close proximity to another, and the flow of current in the two conductors is in opposite directions. When this happens, current will tend to concentrate very close to the region where the conductors are close. This becomes important when considering the "V" (Figure 3.26) where the two tube or pipe walls come together to form the weld. Current flows around this "V" and stays very close to the surface due to the skin effect. Since at any given time, current will be flowing in opposite directions on the two sides of the "V," the proximity effect further assists in concentrating the current at the surfaces, assuming the "V" angle is not too large. However, as the figure indicates, if the "V" angle is too small the sides of the "V" will become too close, and arcing may occur. The concentrated current heats the material to its forging temperature where it is welded by a forging

Figure 3.26 The shape of the "V" where the
two walls of the tube or pipe come together to
form the weld is very important.

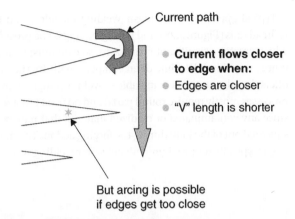

Current path

● **Current flows closer
to edge when:**

● Edges are closer

● "V" length is shorter

But arcing is possible
if edges get too close

action as the two edges are forced together. As a result, all High Frequency welds are intended to
be solid-state welds. The ability of this process to concentrate the current (and therefore the heat)
so effectively is why it is considered the most efficient of all welding processes.

3.6 Flash Welding

Flash Welding (Figure 3.27) is considered a resistance welding process, but the mechanism of heat
generation is very different from other resistance welding processes. Flash Welding relies on a
"flashing" action that occurs when voltage is applied as the two parts to be welded are brought into
close contact. The flashing action is the result of extreme current densities at localized contact
points as the parts come into contact. These tiny regions of extreme current density create rapid
melting, violent expulsion, and vaporization. The flashing action continues for some period of time
to sufficiently heat the surrounding material to its forging temperature. Once sufficient heating
has occurred, an upset force is applied which squeezes both molten and plasticized material out of
the weld region into what is called the flash. Because the primary welding action is caused by the
forging of the plasticized material, Flash Welding is often considered a solid-state welding process
(even though molten material is generated during flashing).

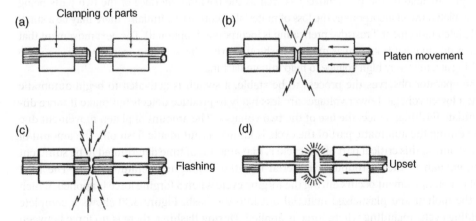

Figure 3.27 The primary steps of flash welding beginning with step (a). (*Source:* Reproduced by permission
of American Welding Society, ©*Welding Handbook*).

Typical applications for Flash Welding include chain links, automotive wheels, window frames, railroad rails (Figure 3.28), and jet engine rings. The process results in violent ejection of molten metal, which creates a significant fire hazard, so care must be taken to keep flammable materials a safe distance away. Flash Welding often competes with Inertia and Continuous Drive Friction Welding, and offers the advantages of being able to weld rectangular cross sections (whereas these friction welding processes mostly require round parts) and no need to rotate the parts. No surface preparation is needed since any contaminated or oxidized material that may be present is expelled during flashing and/or squeezed out of the joint during the forging action. The material squeezed out is referred to as the flash and is typically removed immediately after welding when it is still hot and easy to shear off.

Figure 3.28 Flash welding of railroad rails. (*Source:* Reproduced by permission of American Welding Society, ©*Welding Handbook*).

Figure 3.29 represents a complete Flash Welding cycle that tracks the movement of the platen (only one of the platens holding the parts moves) as well as the flashing interface of the two parts being welded. The platen travel incorporates the loss of material in both parts. Initially, there may be a small amount of platen movement if resistive preheating is incorporated (optional). The next movement that takes place is due to what is known as manual flash-off. During this step, the operator may choose to begin flashing at a relatively high voltage to help initiate flashing.

Once the operator observes the process to be stable, a switch is activated to begin automatic flashing at a lower voltage. Lower voltages are less likely to produce defects but make it more difficult to initiate flashing; hence, the use of the two voltages. The amount of platen movement due to flashing during the automatic part of the cycle is known as automatic flash-off. The amount of time needed during this critical part of the cycle is that amount of time determined to be sufficient for raising enough of the surrounding material into the forging temperature range. The final amount of platen movement occurs during the forging cycle when a forging force is applied, which squeezes the molten and plasticized material out into the flash. Figure 3.30 shows a complete Flash Welding cycle, including where force is applied. During flashing, there is no force between the parts, only sufficient movement of the platen to maintain flashing.

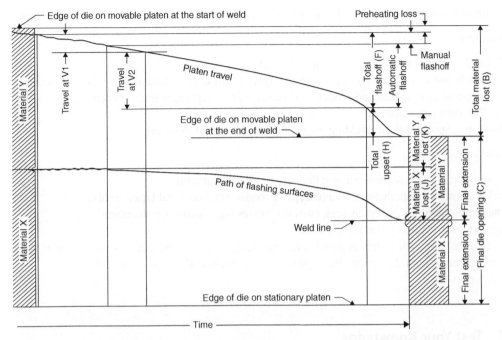

Figure 3.29 A complete flash welding cycle. (*Source:* Reproduced by permission of American Welding Society, ©*Welding Handbook*).

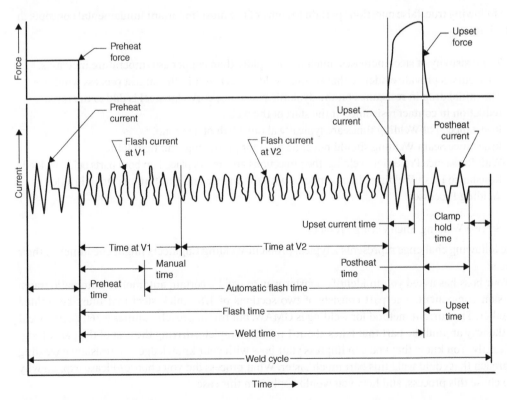

Figure 3.30 A complete flash welding cycle including the application of force. (*Source:* Reproduced by permission of American Welding Society, ©*Welding Handbook*).

In summary, most resistance welding processes offer the following advantages and limitations:

Advantages:

- Can weld most metals, but works best with steel
- Extremely fast welding speeds are possible (a typical spot weld is produced in 1/5th of a second)
- Very good for automation and production because of the "self-clamping" aspect of the electrodes
- No filler materials required
- Resistance Spot and Seam Welding are ideal for welding of thin sheets

Limitations:

- Equipment is much more expensive than arc welding equipment
- Welds cannot be visually inspected (except for Resistance Flash and Upset welds)
- The requirement for extremely high currents creates high power line demands
- In most cases, equipment is not portable
- Mechanical properties such as tensile and fatigue of welds made from processes such as spot welding can be poor due to the sharp geometrical features at the edge of the weld
- Electrode wear

3.7 Test Your Knowledge

I) Fundamental Concepts—True/False

The following true/false questions pertain to some of the most important fundamental concepts in this chapter:

1) The resistivity of steel increases much more rapidly than copper as temperature increases.
2) Lobe curves provide evidence that shorter weld times result in the largest process windows.
3) The initial drop in resistance on the dynamic resistance curve for steel is due to the rapid reduction in contact resistance at the start of the weld.
4) Resistance Spot Welding times are typically about 1/5th of a second.
5) Resistance Seam Welding should never be used for producing leak tight joints.
6) With Resistance Projection welding, the projections are always placed on both parts being welded.
7) Resistance Flash Welding often competes with the Inertia and Continuous Drive Friction Welding Processes.

II) Solve a Welding Engineering Problem

The following challenge represents a typical problem Welding Engineers might encounter in their career:

Your boss has asked you to identify a faster way to weld a certain automotive automatic transmission component. The part consists of two sections of 1/8″ thick steel that must be welded together. The current method for welding is GMAW, but this approach requires a 360 degree weld all the way around the part that takes about 1 min to complete, driving the cost of the part up significantly. You know that you can impress your boss with your knowledge of a resistance welding approach that could weld this part much faster. What process did you choose? Please explain why you chose this process, and how you would apply it in this case.

Recommended Reading for Further Information

ASM Handbook, Tenth Edition, Volume 6—"Welding, Brazing, and Soldering," ASM International, 1993.

AWS Welding Handbook, Ninth Edition, Volume 3—"Welding Processes, Part 2," American Welding Society, 2007.

Resistance Spot Welding—Fundamentals and Applications for the Automotive Industry, Morgan & Claypool, 2018.

Resistance Welding Manual, Revised Fourth Edition, Resistance Welder Manufacturers' Association, 2003.

Resistance Welding—Fundamentals and Applications, CRC Press, 2006.

4

Solid-State Welding Processes

4.1 Fundamentals and Principles of Solid-State Welding

Solid-state welding refers to a family of processes that produce welds without the requirement for molten metal. Solid-state welding theory emphasizes that the driving force for two pieces of metal to spontaneously weld (or form a metallic bond) to each other exists if the barriers (oxides, contaminants, surface roughness) to welding can be eliminated. All solid-state welding processes are based on this concept, and use some combination of heat, pressure, and time to overcome the barriers. Approaches include Resistance, Friction, Diffusion, Explosion, and Ultrasonic Welding.

Since there is no melting, there is no chance for the formation of defects that are only associated with fusion welding processes such as porosity, slag inclusions, and solidification cracking. Solid-state welding processes also do not require filler materials, and in some cases, can be quite effective at welding dissimilar metals that cannot be welded with conventional processes due to metallurgical incompatibilities. The equipment is typically very expensive, and some processes involve significant preparation time of the parts to be welded. Most of these processes are limited to certain joint designs, and some of them are not conducive to a production environment. Nondestructive testing methods do not always work well with solid-state welding processes because of the difficulties associated with distinguishing a true metallurgical bond when there is no solidification.

4.1.1 Solid-State Welding Theory

Materials are made up of one of three common types of atomic bonds: ionic, covalent, or metallic. Although the differences in the types of atomic bonds pertain to how valence electrons are shared, the result is the characteristics and properties of any material relate directly to how the material's atoms are bound. Metallic bonds are unique in that the valence electrons are not bound to a single atom but are free to drift as a "cloud" of electrons. This type of bonding produces many of the characteristics metals are known for, such as electrical conductivity. The forces holding the metal atoms together are the attractive forces between the negatively charged electron cloud and the positively charged nuclei or ions. This suggests that all metals should spontaneously weld together if their atoms are simply brought close enough together. Solid-state welding theory is based on the concept of adhesion, which refers to the spontaneous welding that should occur due to the atomic attraction forces if two metals are brought into close enough contact.

As shown in Figure 4.1, the force of attraction between the ions and the electrons needed for adhesion only occurs when the interatomic spacing is less than 10 Å, although this distance is a

Welding Engineering: An Introduction, Second Edition. David H. Phillips.
© 2023 John Wiley & Sons, Inc. Published 2023 by John Wiley & Sons, Inc.
Companion Website: www.wiley.com/go/Phillips/WeldingEngineeringIntroduction

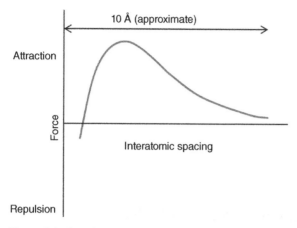

Figure 4.1 In order to achieve spontaneous bonding between two metals, the interatomic spacing must be very close.

function of the metal type. To achieve such closeness between two pieces of metal would require a near perfectly smooth and perfectly clean surface. No common industrial process exists that can create a surface roughness (distance between asperity peaks and valleys) less than 10 Å. Even when a metal is polished to a mirror finish, its microscopic surface roughness is far greater than 10 Å. In addition, metals are also known to rapidly form surface oxides that further inhibit the ability to achieve interatomic closeness between two metal surfaces. Oils and other contaminants are often present as well. In summary, to achieve adhesion or metallic bonding (welding) between two metals without melting, these barriers of surface roughness, oxides, and other contaminants must be overcome. Overcoming these barriers is the goal of all solid-state welding processes.

4.1.2 Roll Bonding Theory

Roll Bonding theory describes the common goal of all solid-state welding processes—to overcome the barriers to bonding and achieve metal-to-metal contact. Roll Bonding (or Welding) is a solid-state welding process used in a variety of applications from clad refrigerator components to coins such as the US quarter. It also provides for a good way to explain the concept of creating nascent surfaces.

Figure 4.2 shows what happens when two pieces of metal are forced through rigid rollers that are separated by a distance that is less than the total thickness of the two metal pieces. The metal

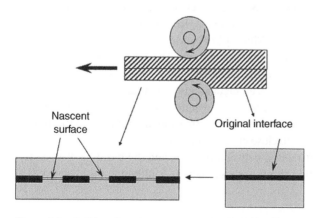

Figure 4.2 Roll bonding theory is based on the creation of nascent surfaces.

thicknesses will be reduced, while at the same time the lengths of the pieces will be increased. The process of increasing their length will automatically increase the length of the interfacial region between the two pieces. In effect, the interfacial region is stretched and squeezed, causing oxides to break apart and surface asperities to collapse. What remains between these regions representing the original surfaces are what is known as nascent surfaces, which are fresh metal surfaces that are in direct contact with each other. These surfaces will begin to spontaneously weld, and creation of a sufficient amount of these surfaces is required to achieve a strong solid-state weld.

The method for creating nascent surfaces varies depending on the type of solid-state welding process. Some examples include Inertia and Continuous Drive Friction Welding processes which rely on the generation of heat to soften the metal, allowing for plastic flow and forging action. Diffusion Welding requires significant joint preparation to remove oxides and contaminants, followed by long times with moderate heat and pressure. Explosion Welding relies primarily on tremendous pressure and the creation of a type of cleaning action known as a "jet."

4.2 Friction Welding Processes

This family of processes relies on frictional heating and significant plastic deformation or forging action to overcome the barriers to solid-state welding. There are a variety of ways to create frictional heating, but the friction welding processes which use the rotation of one part against another are Inertia and Continuous (or Direct) Drive Friction Welding. These are the most common of the friction welding processes and are ideal for round components, bars, or pipes.

In both Inertia and Continuous Drive Friction Welding, one part is rotated at high speed while the other part remains stationary (Figure 4.3). The nonrotating part is then brought into contact with the rotating part with significant force, creating frictional heating that softens (reduces the yield strength) the material at and near the joint. As the process continues, frictional heating diminishes, and heating due to plastic deformation (Figure 4.4) begins to dominate as the softened material is forged out of the joint area. Plastic deformation heating is the same type of heating that occurs when a metal wire is rapidly bent back and forth. Following enough time to properly heat the parts, a higher upset force may be applied near the end of the weld cycle which squeezes the remaining softened hot metal and any contaminants out into the "flash." The flash is typically removed immediately after welding by a machining operation while it is still hot and easy to remove.

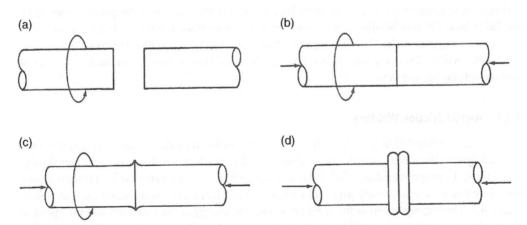

Figure 4.3 Basic steps of both inertia and continuous drive friction welding beginning with step (a). (*Source:* Reproduced by permission of American Welding Society, ©*Welding Handbook*).

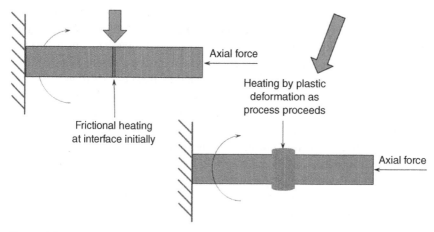

Axial force

Heating by plastic
deformation as
process proceeds

Frictional heating
at interface initially

Axial force

Figure 4.4 Initial heating is due to friction, but then plastic deformation heating dominates.

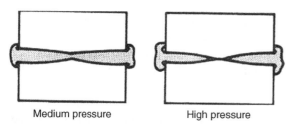

Medium pressure   High pressure

Figure 4.5 Excessive pressure may result in insufficient heating in the center of the weld. (*Source:* Reproduced by permission of American Welding Society, ©*Welding Handbook*).

Since the initial heating is due to friction heating, it is obvious that greater amounts of heating will occur along the outer diameter of the parts being heated since the radial velocity will be the greatest there. Proper heating through the full thickness of the part therefore depends on an amount of time that is sufficient for heat to flow from the heated material in the outer diameter to the inner diameter. Parameters that are adjusted in such a way that do not permit proper heating of the material along the inner diameter may result in lack of bonding in the center. For example, pressures that are too great may prematurely squeeze out heated material along the outer diameter before the region in the center is sufficiently heated (Figure 4.5). This nonuniform heating is also what results in the characteristic hourglass shape of these welds.

Other approaches to friction welding include a method that allows rectangular shapes to be welded (Linear Friction Welding), and a method using an additional pin tool that can create friction welds using conventional joint designs such as butt joints (Friction Stir Welding: FSW). These approaches will be discussed later in this chapter. But in all cases, the fundamental principles of friction welding are the same.

4.2.1 Inertia Friction Welding

With Inertia Friction Welding (Figure 4.6), a flywheel is used to provide the energy required to spin one part against the other to create the heating needed to produce a weld. A motor brings the flywheel up to the proper speed and then a clutch releases it, allowing it to spin free. It is then moved into place against the stationary part, and a force is applied. A typical weld sequence is shown in Figure 4.7. The forging action at the weld consumes the energy of the flywheel, and the speed of the flywheel decreases as the weld region begins to cool and regain strength. An additional forging force may be applied at the end of the cycle to create the proper amount of upset. Two very

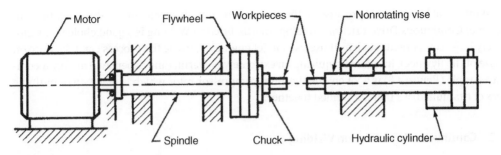

Figure 4.6 Inertia friction welding. (*Source:* Reproduced by permission of American Welding Society, ©*Welding Handbook*).

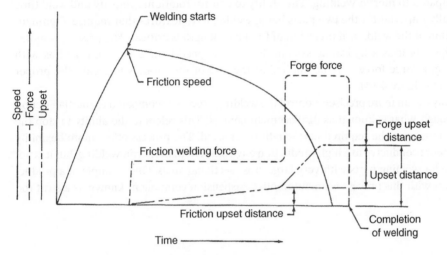

Figure 4.7 Typical inertia friction welding sequence. (*Source:* Reproduced by permission of American Welding Society, ©*Welding Handbook*).

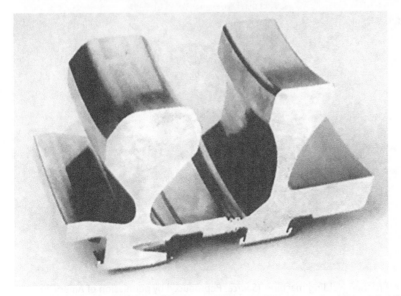

Figure 4.8 Inertia welded section of titanium compressor rotor for a jet engine. (*Source:* Reproduced by permission of American Welding Society, ©*Welding Handbook*).

important variables with this process are the size and the initial velocity of the flywheel. In comparison to Continuous Drive Friction Welding, Inertia Friction Welding is a good choice for welding very large components such as those used in jet engines, because the process can rely more on the size of the flywheel, instead of requiring an extremely powerful (and expensive) motor. A common example (Figure 4.8) is a jet engine titanium compressor rotor which is made up of multiple stages of titanium disks that are welded together.

4.2.2 Continuous Drive Friction Welding

Continuous Drive Friction Welding (Figure 4.9), also referred to as Direct Drive Friction Welding, offers more precise control of the rotating part and no requirement to change flywheels as compared to Inertia Welding. The ability to control rotational velocity and weld time can be especially important if the two parts being welded have features that require alignment upon completion of the weld, or if precision of final part length is critical. The process is motor driven and typically uses a hydraulic system. Frictional speed is held constant, and as with Inertia Welding, a forge force may be applied at the end of the process to create the proper amount of upset (Figure 4.10).

One of the ways it can more precisely control the welding process as compared to Inertia Friction welding is through what is known as displacement control. This refers to the ability to stop the welding cycle as soon as the required part length is achieved. The process offers more flexibility and is usually more conducive to a high production environment than Inertia Welding, but it is not the best choice for welding parts with very large cross-sectional areas. One example of a production welded part with this process is an automotive suspension component known as a strut rod (Figure 4.11).

Figure 4.9 Continuous drive friction welding machine. (*Source:* Reproduced by permission of American Welding Society, ©*Welding Handbook*).

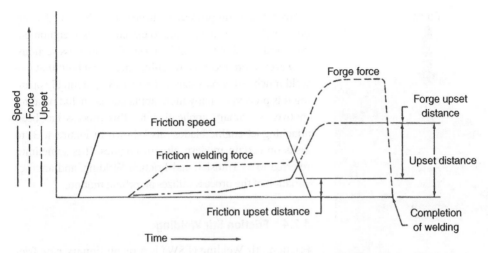

Figure 4.10 Typical continuous drive friction welding sequence. (*Source:* Reproduced by permission of American Welding Society, ©*Welding Handbook*).

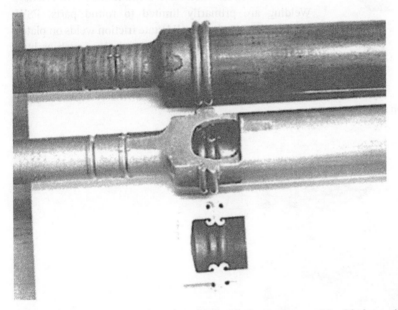

Figure 4.11 Automotive strut rods welded with the continuous drive friction welding process. (*Source:* Reproduced by permission of American Welding Society, ©*Welding Handbook*).

4.2.3 Linear Friction Welding

The Linear Friction Welding process (Figure 4.12) utilizes straight-line or linear oscillation to produce relative motion between two parts. This motion is combined with forge loading normal to the oscillation surface to create the frictional heating required for welding. The linear motion allows for a wide variety of part shapes (including rectangular) to be friction welded. Surface velocity and forge force can be used to vary the rate of power input needed to accommodate different part sizes.

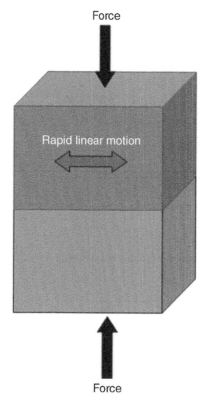

Force

Rapid linear motion

Force

Figure 4.12 Linear friction welding.

Historically, this process has been primarily limited to aircraft engine applications due to extremely high equipment costs and the ability to weld only small weld cross-sections. However, more recently, machines are being built that can weld much larger part sizes (Figure 4.13), potentially opening this process to many more applications, including automotive. A primary application for this process has been attaching jet engine blades (also shown in Figure 4.13) to disks on certain modern aircraft engines. It is sometimes referred to as Translational Friction Welding, and another variation of the process utilizes an orbital motion.

4.2.4 Friction Stir Welding

Friction Stir Welding (FSW) is a revolutionary new friction welding process that was developed in the early 1990s by the British Welding Institute (TWI). Whereas processes such as Inertia and Continuous Drive Friction Welding are primarily limited to round parts, FSW (Figure 4.14) produces solid-state friction welds on plates using conventional arc weld joint designs such as butt joints. FSW relies on the frictional heat created when a specially designed nonconsumable pin tool is rotated along the joint against the top of the two pieces being

Figure 4.13 Large-scale linear friction welder. (*Source:* Manufacturing Technology, Inc).

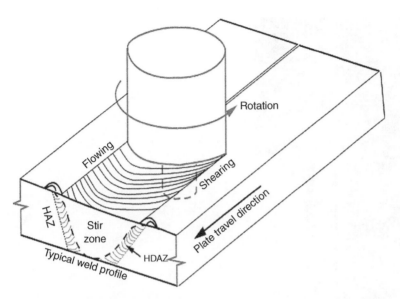

Figure 4.14 Friction stir welding.

welded. As the process begins, frictional heating softens (or plasticizes) the metal at the joint, facilitating a stirring action of plasticized metal. Further heat that is generated by the plastic deformation becomes the main source of heating as the weld progresses.

The pin tool for FSW consists of a shoulder and pin. The shoulder rests on the surface of the plates and provides most of the frictional heating. The pin partially penetrates the joint of the plates being welded. Within the stir zone itself, the plasticized hot metal flows around the pin to form a weld. The primary purpose of the pin is to control the stirring action. A wide variety of pin tool designs and materials have been studied, including the use of tungsten-based materials, threaded pins, and cupped shoulders. A slight push angle (<5°) relative to the travel direction is sometimes used, and travel speeds are much slower than typical arc welding speeds. FSW machines tend to be large and are built with very rigid construction (Figure 4.15). Applications for this process are growing; a famous application is the welded seam of the aluminum alloy main fuel tank on the Space Shuttle. Other applications include the seams along the tubular structures used for manufacturing launch vehicles (rockets) and some automotive sheet connections.

As shown in Figure 4.16, three distinct regions can be identified in a weld produced with Friction Stir Welding. The stir zone represents the region where there is considerable metal flow or stirring due to the tool rotation. The heating combined with plastic deformation continually creates a very fine grain structure, a process known as dynamic recrystallization. This fine grain structure associated with the stir zone exhibits excellent mechanical properties. As a result, one of the major advantages of this process is improved as-welded tensile properties of aluminum weldments as compared to those produced with arc welding processes. The stir zone microstructure is likely to vary from one side to the other since the relative velocity of the stirring material will be greater in the direction of weld travel. The heat and deformation-affected zone (HDAZ) is a narrow region surrounding the primary stir zone, which reveals small levels of deformation that occurs at the elevated welding temperatures. The heat-affected zone (HAZ) is only influenced by the heat flow out of the weld. The metallurgical reactions that occur in this region are essentially identical to those in a fusion weld.

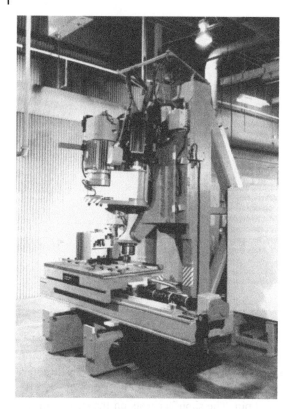

Figure 4.15 Typical friction stir welding machine. (*Source:* Reproduced by permission of American Welding Society, ©*Welding Handbook*).

Figure 4.16 Regions of a friction stir weld.

To successfully weld a material with this process, it is necessary for the material to flow at an elevated temperature. Therefore, flow stress as a function of temperature can be used to determine the Friction Stir Welding characteristics of a material. In general, the more rapidly flow stress drops with increasing temperature for a given metal the more easily it can be welded by this process. In the example in Figure 4.17, Alloy A would be expected to be much easier to weld than Alloy B because it exhibits a lower flow stress in the stir zone temperature range. Higher stirring temperatures also affect tool wear, which is a major consideration with this process.

When welding aluminum alloys, temperatures above 450°C (840°F) are needed for sufficient metal flow to occur. For steels, temperatures in excess of 1200°C (2190°F) are required. In the early years of development of FSW, the use of tool steels for the pin tool limited its application to metals like aluminum. However, with the recent developments of higher temperature pin tools, metals such as steel, titanium, and even nickel are now successfully being welded with reasonable tool life.

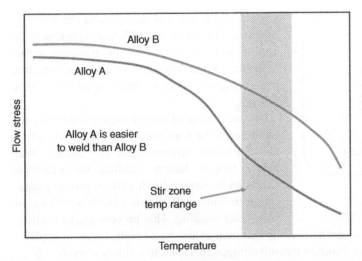

Figure 4.17 Flow stress at high temperature of the metal being welding is a very important consideration with friction stir welding.

In summary, the advantages and limitations of all friction welding processes are as follows:

Advantages:

- No melting means no chance for solidification-related defects
- Filler materials are not needed
- Very few process variables result in a very repeatable process
- Can be used in a production environment (this mainly applies to the Continuous Drive Friction Welding process)
- Fine grain structure of friction welds typically exhibits excellent mechanical properties relative to the base metal, especially when welding aluminum
- No special joint preparation or manual welding skill required

Limitations:

- Equipment is very expensive
- Limited joint designs, and in the case of Continuous Drive and Inertia Friction Welding, parts must be round and symmetric
- Friction Stir Welding is very slow, and therefore, not conducive to high-speed production
- Part size limitations with Linear Friction Welding

4.3 Other Solid-State Welding Processes

4.3.1 Diffusion Welding

Diffusion Welding (Figure 4.18) is a solid-state welding process that relies on diffusion to create a weld through a combination of heat and pressure. It generally requires long times (as much as 1 hour) at elevated temperatures and is usually conducted in a vacuum or protective

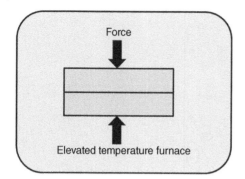

Figure 4.18   Diffusion welding.

atmosphere to prevent oxidation at the welding interface. Pressure of sufficient magnitude is applied to cause some local deformation at the interface. A machine known as a vacuum hot press (Figure 4.19) is often used with this process.

One advantage of this process is minimal degradation to the base metal microstructure since the maximum temperatures are usually well below typical fusion welding temperatures. Diffusion Welding relies highly on surface preparation to remove the barriers (such as oxides) to solid-state welding. This process works particularly well with titanium which readily dissolves its own oxide. As a result, common applications for Diffusion Welding are titanium aircraft components used in military aircraft.

To get sufficient diffusion to create a weld, intimate contact between the parts must be achieved. Prior to welding, surface preparation techniques that typically combine metal machining or grinding with chemical etching create smooth surfaces and remove most of the oxides and contaminants. But as discussed previously, no industrial machining process can achieve perfect smoothness and cleanliness, so when the parts are brought together there are still localized asperity peaks and valleys, and some oxides that must be overcome (Figure 4.20). The application of heat reduces the yield strength of the material, and a moderate amount of general pressure will create local pressures high enough to plastically deform the asperity peaks and create the required intimate contact. Higher temperatures will reduce welding times, but the temperatures should not be too high to affect the microstructure and potentially degrade the mechanical properties of the part being

Figure 4.19   Vacuum hot press. (*Source:* Reproduced by permission of American Welding Society, ©*Welding Handbook*).

welded. And a general rule of thumb is to choose a pressure that is slightly below the yield stress of the material at the welding temperature.

AWS considers the entire Diffusion Welding process to consist of three stages after the initial contact is made (Figure 4.21). As the asperities collapse and intimate contact is achieved, an interfacial boundary forms. The interfacial boundary is much like a grain boundary, which over the course of the Diffusion Welding cycle, lowers its free energy by migrating into the surrounding microstructure. When the weld is complete, there will usually be minimal evidence of a bond line, although the creation of pores is a possibility. Some metals such as nickel are more difficult to diffusion weld, so a similar process called Diffusion Brazing is often used. Diffusion Brazing uses a filler metal that melts at temperatures below the base metal melting temperature, but later diffuses away leaving little evidence of its existence.

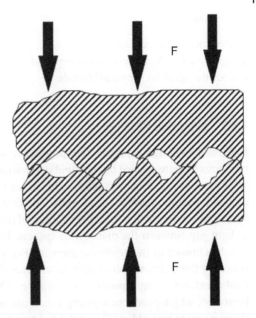

Figure 4.20 Pressure combined with lowering of the yield strength due to heating collapses the asperity peaks and creates intimate contact (F refers to force).

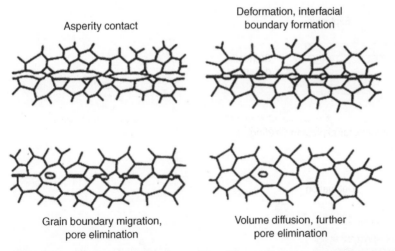

Figure 4.21 The stages of diffusion welding. (*Source:* Reproduced by permission of American Welding Society, ©*Welding Handbook*).

In summary, the advantages and disadvantages of Diffusion Welding are as follows:

Advantages:

- Minimal degradation to the base metal
- No distortion or deformation

Limitations:

- Extremely long weld times
- Significant surface preparation required
- Does not work well with all metals

4.3.2 Explosion Welding

Explosion Welding is a process that produces a solid-state weld through the application of extreme pressures that are created by the rapid detonation of an explosive material. It typically involves the welding of two plates and is often used to join metallurgically incompatible materials or for cladding of pipe and pressure vessel walls. The first step of the process is to remove oxides and other contaminants (typically through grinding) of the surfaces to be welded followed by the creation of a slight gap between the plates with spacers. Explosive material is then placed on the top plate, which is known as the prime component (Figure 4.22). The detonation is typically started from a corner of this plate, which in turn accelerates the prime component down onto the base component at extremely high speeds on the order of 300 m/s. As the two plates collide, extreme pressure is created, which produces a jetting action ahead of the weld that provides additional cleaning of the surfaces to be welded (Figure 4.23). Important process variables include the standoff distance (gap between the plates), the collision angle between the prime and base components, and the velocity of the detonation.

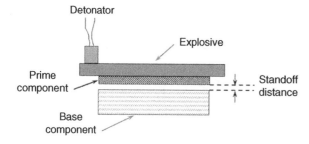

Figure 4.22 Typical arrangement for explosion welding.

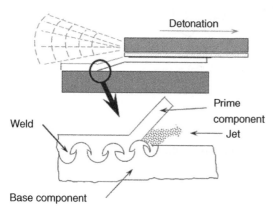

Figure 4.23 Explosion welding concept.

The weld forms almost instantaneously, thereby suppressing most metallurgical reactions such as the formation of brittle intermetallic phases. The weld interface typically takes on a characteristic wavy appearance. Explosion Welding relies primarily on extreme pressure and is therefore capable of producing welds with almost no heating (although localized melt pockets are often observed at the interface, likely due to regions of extreme plastic deformation). Because of the miniscule amount of heat generation, it is the solid-state welding process of choice when welding dissimilar metal combinations that cannot be welded by any other method. As shown in Figure 4.24, Explosion Welding typically involves six distinct manufacturing steps: (1) material inspection, (2) grinding of surfaces, (3) assembly, (4) explosion, (5) flattening and cutting, and (6) ultrasonic inspection to identify the welded region (the outer edges of the plates typically do not get welded and are removed).

Cladding is a common application of Explosion Welding. For example, in extreme corrosive environments, materials such as nickel, titanium, or tantalum which resist corrosion can be clad to steel, which is much less expensive. Because corrosion resistance can be achieved with a relatively thin layer of the corrosion-resistant metal, the required strength of the component can be attained through sufficient wall thickness of the inexpensive steel. In this way, a corrosion-resistant component can be produced that is relatively inexpensive. The clad plates are then formed and welded into place to create a variety of fabrications, including pressure vessels and reactor components. Explosion Welding can also be used to produce dissimilar metal "transition joints,"

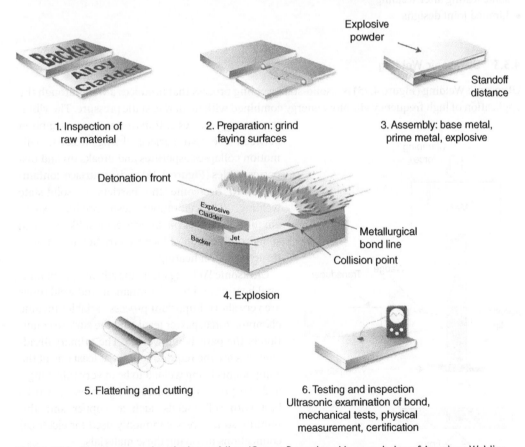

Figure 4.24 The six steps of explosion welding. (*Source:* Reproduced by permission of American Welding Society, ©*Welding Handbook*).

which are later welded into place using conventional welding processes. An example is an aluminum-to-steel transition joint used for some shipbuilding applications.

The process is not limited to flat plates. A common application involving a round part is the welding of titanium tubes to the inside of holes drilled into steel plates (known as tube sheets) in heat exchanger applications. This approach involves placing explosives inside each tube, which upon detonation, create an outward pressure on the tubes producing the weld between the outside of the tube and the inner wall of the tube sheet. A weld between titanium and steel would be nearly impossible to produce with a conventional process such as arc welding because of the brittle intermetallic formation that would occur in the presence of high heat and large amounts of molten metal.

In summary, the advantages and limitations of Explosion Welding are as follows:

Advantages:

- Extremely low heat input process allows for welding of nearly any dissimilar metal combination
- Allows for the creation of relatively inexpensive fabrications that are very resistant to corrosion

Limitations:

- Not conducive to a production environment
- Requires many steps, including surface preparation prior to welding, and flattening and ultrasonic testing after welding
- Limited joint designs

4.3.3 Ultrasonic Welding

Ultrasonic Welding (Figure 4.25) is a solid-state welding process that produces a weld through the application of high frequency vibratory energy combined with moderate static pressure. The vibratory energy creates a transverse shearing motion between the two surfaces being welded. This motion collapses asperities and breaks up and disperses oxides (Figure 4.26) and/or surface contaminants to overcome the barriers to solid-state welding. The vibrations also produce small amounts of heat at the interface, most likely due to some combination of localized frictional and plastic deformation heating.

Ultrasonic Welding offers the advantages of minimal heating and part deformation, and weld times are very short. Important process variables include clamping force, power level, and the surface condition of the parts being welded. The primary disadvantages are the requirement for at least one of the components being welded to be of very thin gauge, and the process is limited to lap joints. It works best with soft metals such as copper and aluminum, so it is very commonly used for electrical connections involving these materials.

The two most common types of ultrasonic welding systems are the lateral-drive and wedge-reed

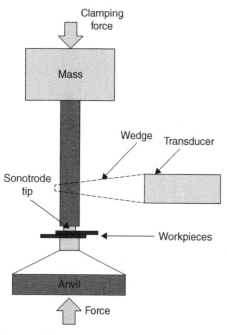

Figure 4.25 Ultrasonic welding.

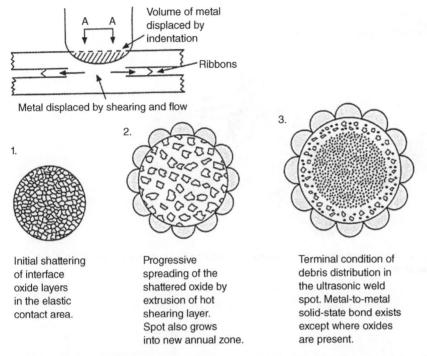

Figure 4.26 Ultrasonic welding involves a shearing action that first shatters and then disperses oxides. (*Source:* Reproduced by permission of American Welding Society, ©*Welding Handbook*).

systems. Figure 4.25 shows a wedge-reed machine. Both systems consist of three common components: a power source, an ultrasonic transducer, and a horn. The power source converts the line frequency to the required ultrasonic welding frequency. This frequency is then converted to mechanical vibrations of the same frequency by the transducer through the use of piezoelectric material, which expands and contracts when exposed to the alternating voltage supplied by the power source. The wedge (or horn) amplifies the vibrations that are then transmitted to the part by the sonotrode. The high frequency vibrations (20 kHz is typical) are relatively low in amplitude (20 μm is typical). A typical Ultrasonic Welding machine which utilizes a lateral drive system is shown in Figure 4.27.

As mentioned, a wide variety of electrical connections such as wire splices and cable terminations are produced with Ultrasonic Welding in various industry sectors, including automotive. In fact, machines specifically designed for this purpose may be referred to as wire splicing machines. Figure 4.28 shows one of these machines as well as an example of a typical wire splice weld that would be produced with this type of machine.

Although most Ultrasonic Welding machines rely on pneumatic pressure to generate the force needed for welding, recent development in machine technology involves utilizing servo motors similar to modern Resistance Spot Welding machines discussed earlier. The potential benefits of servo motor-driven machines are comparable to the Resistance Spot Welding benefits. Servo motors offer much more control which makes it possible to weld in multiple steps and vary the force with each step. This ability provides for much more flexibility when encountering difficult welding applications. They also allow for closed-loop feedback control as the weld is being made, which can provide for much more consistent welds and better regulation of the weld height and area.

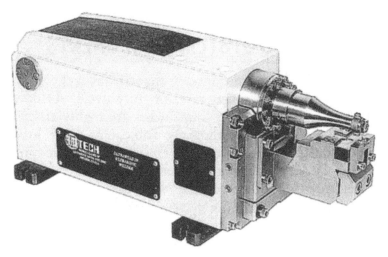

Figure 4.27 Typical ultrasonic welding machine. (*Source:* Reproduced by permission of American Welding Society, ©*Welding Handbook*).

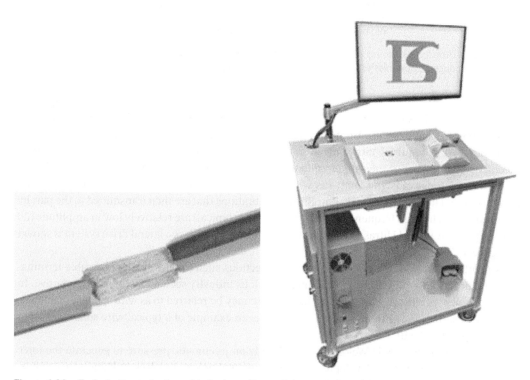

Figure 4.28 Typical ultrasonically welded wire splice and the type of machine used to produce this weld. (*Source:* Tech-Sonic).

Figure 4.29 Closed-loop control ultrasonic spot welding machine. (*Source:* Tech-Sonic).

These machines are referred to as closed-loop control machines, and a spot welding version is shown in Figure 4.29. Such advanced capability could become increasingly important in the automotive industry considering the rapid worldwide push toward electric cars. These cars utilize batteries that are made up of hundreds of individual cells that must be interconnected with a welding process, and Ultrasonic Welding offers many benefits since it works so well with copper and aluminum, and there is no melting involved.

Ultrasonic Welding is also a very popular method for welding plastics (see Chapter 6), but the approach is very different. The machines designed for plastic welding induce vibrations that are perpendicular to the lap joint as opposed to parallel when welding metals. Also, Ultrasonic Welding of plastics involves melting of the plastic, which is not the case with metals.

In summary, the advantages and limitations of Ultrasonic Welding are as follows:

Advantages:

- Very low heat input and minimal part distortion (other than surface marking)
- Fast welding speeds
- Ideal for making electrical connections

Limitations:

- At least one of the parts being welded must be very thin
- Mostly limited to soft metals
- Limited to lap joints

4.4 Test Your Knowledge

I) Fundamental Concepts—True/False

The following true/false questions pertain to some of the most important fundamental concepts in this chapter:

1) One of the unique aspects of metallic bonds is that the electrons are not bound to a single atom.
2) Spontaneous bonding (welding) of metals can easily be demonstrated by simply polishing two pieces of metal and placing them in contact with each other.
3) The goal of all solid-state welding processes is to achieve metal-to-metal contact through the creation of nascent surfaces.
4) The heat generated in all friction welding processes is entirely derived from the frictional heating between the two pieces being welded.
5) Inertia Friction Welding offers much more control than Continuous Drive Friction Welding.
6) A major benefit of Friction Stir Welding is fast welding speeds.
7) Diffusion Welding requires very long weld times.
8) Explosion Welding relies on the generation of large amounts of heat to produce the weld.
9) Ultrasonic Welding works well with all metals.

II) Solve a Welding Engineering Problem

The following challenge represents a typical problem Welding Engineers might encounter in their career:

Your company is developing a new aerospace component that involves a connection between nickel and titanium. During initial welding trials utilizing the GTAW process, it is quickly discovered that this metal combination cannot be arc welded because of the extremely brittle intermetallics that form. The ability to produce this part has the potential to make the company a lot of money, so your boss has asked you to find a solution. What process do you choose, and why?

Recommended Reading for Further Information

ASM Handbook, Tenth Edition, Volume 6—"Welding, Brazing, and Soldering," ASM International, 1993.
AWS Welding Handbook, Ninth Edition, Volume 3—"Welding Processes, Part 2." American Welding Society, 2007.

5

High-Energy Density Welding Processes

5.1 Fundamentals and Principles of High-Energy Density Welding

High-energy density (HED) welding processes are those that focus the energy needed for welding to an extremely small area. This allows for very low overall welding heat input which results in narrow heat-affected zones and reduced residual stress and distortion. Welding speeds can be very fast as well. The two main processes known for extremely high-energy densities are Laser Beam Welding (Figure 5.1) and Electron Beam Welding.

5.1.1 Power Density

As shown in Figure 5.2, energy densities of focused laser and electron beams can approach and exceed 10^4 kW/cm^2. These energy densities are achieved through a combination of high power and beams that are focused to diameters as small as a human hair (0.05 mm). Plasma Arc Welding, discussed previously, offers greater energy density than conventional arc welding processes, and is sometimes referred to as the "poor man's laser". But as the figure shows, the energy densities possible with Electron Beam and Laser Beam Welding are much greater than any arc welding process. As a result, these two processes are the only two that are commonly categorized as HED welding processes.

HED processes produce weld profiles that exhibit high depth-to-width ratio as compared to other welding processes (Figure 5.3). As a result, much greater thicknesses can be welded in a single pass, especially with Electron Beam Welding. The figure also illustrates the fact that HED processes can produce a weld with minimal heating to the surrounding area as compared to other processes. However, the concentrated heat source and high depth-to-width ratio weld profile produced by these processes renders them to be much more sensitive to imperfect joint fit-up as compared to arc welding processes. Rapid weld cooling rates associated with HED welding also increases the likelihood for forming the brittle martensite phase when welding certain steels.

Laser Beam Welding and Electron Beam Welding processes are used in a wide variety of industrial sectors. In addition to the possibility for high weld speeds, the welds are usually very aesthetically pleasing. Laser Welding is very adaptable to high-speed production, so it is common in the automotive sector (refer Figure 5.1). The ability to precisely locate welds on smaller sensitive components with minimal heat input makes Laser Welding very attractive to the medical equipment industry as well. Electron Beam Welding can weld much thicker components than Laser Welding, but is usually conducted in a vacuum, restricting its use in high-production environments. Electron Beam Welding is popular for jet engine, aerospace, and nuclear components where very

Welding Engineering: An Introduction, Second Edition. David H. Phillips.
© 2023 John Wiley & Sons, Inc. Published 2023 by John Wiley & Sons, Inc.
Companion Website: www.wiley.com/go/Phillips/WeldingEngineeringIntroduction

Figure 5.1 Laser welding in the automotive industry. (*Source:* Reproduced by permission of American Welding Society, ©*Welding Handbook*).

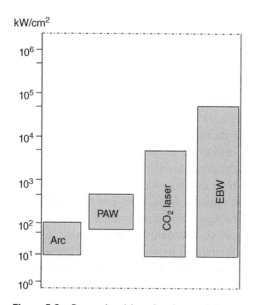

Figure 5.2 Power densities of various welding processes.

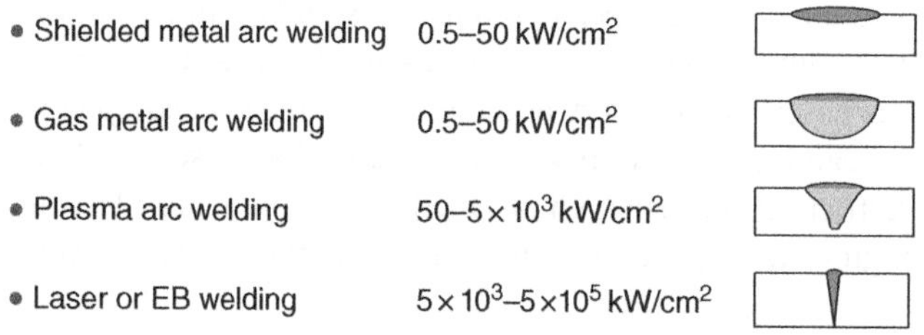

- Shielded metal arc welding $0.5–50 \, kW/cm^2$
- Gas metal arc welding $0.5–50 \, kW/cm^2$
- Plasma arc welding $50–5 \times 10^3 \, kW/cm^2$
- Laser or EB welding $5 \times 10^3–5 \times 10^5 \, kW/cm^2$

Figure 5.3 Comparison of typical weld profiles and the range of energy densities associated with the processes that produce them.

high-quality thick section welding is needed, and high production speeds are not required. But there are also some smaller-scale automotive applications such as the welding of gears.

5.1.2 Keyhole Mode Welding

When welding with HED processes, the laser or electron beam is focused along the joint of the work pieces to be welded. The extreme power density of the beam not only melts the material, but also causes evaporation. As the metal atoms evaporate and exit the surface of the molten metal, forces in the opposite direction of evaporation create a significant localized vapor pressure. This pressure depresses the molten metal surface creating a hole which is known as a keyhole (Figure 5.4). The weld metal solidifies behind the keyhole as it progresses along the joint. This method of welding produces high depth-to-width ratio welds and is the most common approach when using either Laser Beam Welding or Electron Beam Welding. There are some cases where the keyhole mode is not used, an approach known as conductive mode welding. Conductive mode welding produces welds with a wilder weld profile like that of an arc weld and can be especially useful when welding thin materials.

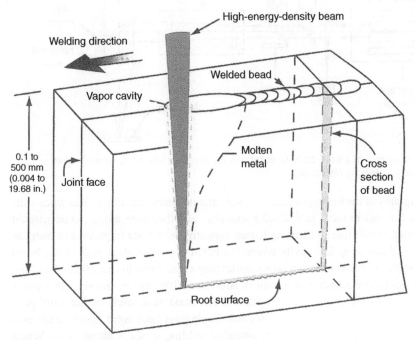

Figure 5.4 Keyhole mode welding. (*Source:* Reproduced by permission of American Welding Society, ©*Welding Handbook*).

5.2 Laser Beam Welding

The word laser is an acronym for "light amplification by stimulated emission of radiation." Lasers produce a special form of light (electromagnetic energy) consisting of photons that are coherent and of the same wavelength. Light of this form can be focused to extremely small diameters allowing for the creation of the high-energy densities used for welding. The laser beam itself is not useful for welding until it is focused by a focusing lens.

A standard Laser Welding machine consists of four primary components: a lasing medium, a pumping source, a lasing cavity, and the focusing optics (Figure 5.5). The lasing medium is a material that can be easily raised to a higher energy level by "pumping" it with an energy source such as a flash lamp or diode. When the atoms (or molecules) of the material are raised to a higher energy level they become unstable since the outer electron shells contain excess electrons. When the electrons later fall back to a lower energy shell, a photon is given off, a process known as spontaneous emission. The stimulated emission step occurs when this photon in turn further excites other atoms or molecules, which then give off other photons, and so on, creating an "avalanche" of photons that are all traveling in phase and are of the same wavelength. This process occurs inside the laser cavity that contains reflective mirrors that bounce the photons back and forth to amplify the power until it is sufficient for lasing.

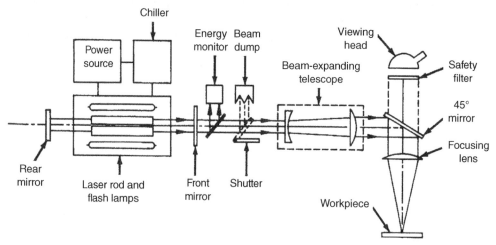

Figure 5.5 Typical components of a solid-state laser system. (*Source:* Reproduced by permission of American Welding Society, ©*Welding Handbook*).

Lasers vary in the quality of the beam produced. A high-quality beam will diffract less when exiting the laser making it easier to focus to a smaller spot size, and therefore, a more concentrated heat source for welding. In addition to focusing lenses, reflective lenses are important to lasers as well since they are used in the optical cavity where the beam is generated, as well in the beam delivery systems for some Lasers. For these reasons, optics play a major role in Laser Beam Welding.

Laser Beam Welding (Figure 5.6) does not require additional filler metal, and shielding gas is optional. The maximum thickness for a full penetration single-pass laser weld (steel) made at a reasonable welding speed is about ¾ in. When the beam impinges on the work piece, it melts and vaporizes metal atoms, some of which are ionized by the intense beam creating a plasma known as a "plume". The plume can have the effect of attenuating the beam which reduces weld penetration. A common approach to minimizing the negative effect of the plume is to deflect it with a flow of gas. Weld spatter onto the focusing lens can also sometimes be a problem, especially when there are contaminants on the surface of the parts being welded. Approaches to minimizing the spatter problem include

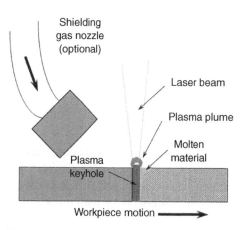

Figure 5.6 Laser beam welding.

choosing a long focal length lens that keeps the lens a safe distance from the weld area, or the use of an air "knife" to protect the lens.

The choice of laser type depends on cost, the type and thickness of the material to be welded, and the required speed and penetration. Lasers are distinguished by the medium used to generate the laser beam, and the wavelength of laser light produced. Although there are many types of lasers, common lasers for welding include the solid-state Nd:YAG, Fiber, Diode, and Disk Lasers, and the gas-based CO_2 Laser. The lasing medium in solid-state lasers are crystals (Nd:YAG and Disk Lasers) or fibers (Fiber Laser) that have material such as neodymium ions added or doped that will "lase" when "pumped" (exposed to a source of energy). The lasing medium in the CO_2 Laser is CO_2 gas. In all cases, "lasing" occurs when the atoms/molecules of the medium are excited to a higher energy state as described previously.

CO_2 Lasers are much less common now than they used to be. They produce wavelengths of 10.6 μm, while the wavelength of most solid-state lasers is in the range of 1.0 μm. A CO_2 Laser is generally less expensive, but the longer wavelength of light does not allow its beam to be delivered through fiber optic cables, which reduces its versatility. Its light is also more reflective, which limits its use with highly reflective metals such as aluminum. The solid-state lasers are generally more compact and require less maintenance than CO_2 Lasers and are more conducive to high-speed production since their beams can be delivered through long lengths of fiber optic cable which can then be attached to launch optics attached to a robot arm. Solid-state lasers such as the Fiber, Diode, and Disk Lasers produce beams of outstanding quality and offer very high electrical efficiencies as compared Nd:YAG and CO_2 Lasers, and therefore, for all these reasons have become the popular lasers of choice for welding. One disadvantage of these lasers is their shorter wavelength represents a dangerous eye hazard resulting in the requirement for special eye protection.

The choices of focus spot size, focus spot location in the joint, and focal length are all important considerations when using Laser Beam Welding. Usually, a small focus size is used for cutting and welding, while a larger focus is used for heat treatment or surface modification. As indicated in Figure 5.7, the location of the beam's focal point can also be varied based on the application. When welding, it is common to locate the focal point somewhere near the center of the joint thickness. Cutting applications benefit from placing the focal point at the bottom of the joint.

In summary, the advantages and limitations of Laser Beam Welding are as follows:

Advantages:

- HED process allows for low overall heat input, which produces small heat-affected zones and minimal degradation to base metal properties, as well as reduced residual stress and distortion
- Fast welding speeds

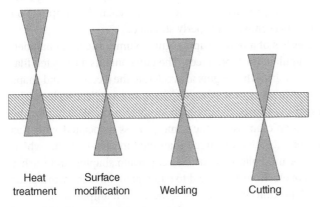

Heat treatment   Surface modification   Welding   Cutting

Figure 5.7 Focusing of the laser beam.

- No filler metal required
- Shielding gas is usually not required, and no need to weld in a vacuum (as compared to EB Welding)
- Relatively thick (3/4 in.) single-pass welds can be made
- Concentrated heat source allows for the creation of extremely small weld sizes needed for small and intricate components
- Solid-state lasers are easily automated
- Since there is no bulky torch as with most arc welding processes, Laser Beam Welding is capable of welding joints that are difficult to access

Limitations:

- Equipment is very expensive
- Portability is usually low
- Requires very tight joint fit-up and accurate positioning of the joint relative to the beam
- Metals that are highly reflective such as aluminum are difficult to weld with some Laser Beam Welding processes
- High weld cooling rates may create brittle microstructures when welding certain steels
- Laser plume may be a problem
- Poor electrical efficiency with some lasers
- Solid-state lasers require special (and expensive) eye protection
- Laser Beam Welding is complex and requires significant training and knowledge

5.3 Electron Beam Welding

Electron Beam Welding is a process that produces a beam of accelerated and focused electrons. The intense heating generated for welding is associated with the kinetic energy created when electrons, which are accelerated to approximately 70% the speed of light, impinge on the part being welded. The volume of electrons creating this heating is surprisingly low, as beam amperages are typically less than 1 amp. The electrons are accelerated through a gun column (Figure 5.8) via the application of extremely high voltages (as high as 200 kV) between the cathode and anode. Electron Beam Welding is usually conducted in a high vacuum, which limits its application in high production environments. However, when welding titanium, a high vacuum can be beneficial since titanium is known to be very sensitive to embrittlement when welding with arc welding processes. Some applications utilize low vacuum and nonvacuum systems, but at the expense of a reduction in energy density of the beam due to beam divergence that occurs when the rapidly moving electrons encounter air molecules. Electron Beams produce dangerous X-rays when they impinge on the part being welded, so the welding chambers must be properly shielded.

An Electron Beam Welding machine consists of a power supply, gun column, vacuum chamber, and appropriate fixturing to hold and manipulate the work piece. The gun contains a tungsten filament that serves as the cathode, a bias cup (or grid) that begins to accelerate the electrons and shape the beam, an anode that further accelerates the electrons, a focusing lens, and a deflecting lens. The process begins when the tungsten is resistively heated to very high temperatures, allowing for the easy liberation of electrons through thermionic emission, the same process associated with Gas Tungsten Arc Welding. The bias cup is more negatively charged than the tungsten cathode, which allows it to both accelerate and shape the beam. Following the bias cup beam shaping and further acceleration through the anode, the beam of electrons is focused to a fine point onto the work piece via a magnetic focusing lens. A magnetic deflection coil allows for some beam manipulation.

By far, Electron Beam Welding can produce the greatest single-pass weld penetrations of any fusion welding processes. Single-pass welding thicknesses of greater than 10″ thick are possible depending on the metal being welded, and the welds are typically very aesthetically pleasing. As with Laser Beam Welding, Electron Beam Welding delivers minimal heat to the part, resulting in small heat-affected zones and minimal degradation to base metal properties, as well as reduced residual stress, and distortion. The depth of focus with electron beams is greater than with laser beams which can provide for more flexibility with part placement. Beam adsorption is much better than lasers (which are susceptible to reflectivity) and the ability to manipulate the beam with the deflection coils offers another advantage.

Electron Beam Welding machines are typically manufactured as either high voltage (i.e., 150 kV) or low voltage (i.e., 60kV) systems. The cost of the machine is a function of both the chamber size and the power of the machine. The high voltage machines are the most expensive and provide the deepest weld penetration. Machines with larger chambers will allow for larger parts to be welded but will be more expensive and require longer pump down times. In addition to the magnetic

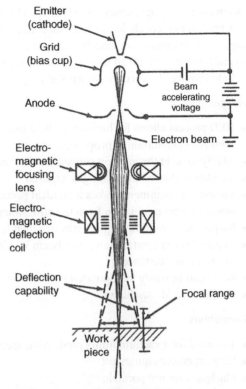

Figure 5.8 Basic components of an electron beam gun. (*Source:* Reproduced by permission of American Welding Society, ©*Welding Handbook*).

deflection coil, many machines offer additional ability to manipulate the part inside the chamber, such as x-y tables and rotary/tilt positioners.

In addition to the aerospace (Figure 5.9) and nuclear sectors, applications for this process exist in a wide variety of industry sectors, including medical, automotive, petrochemical, power generation, and

Figure 5.9 Electron beam welded titanium rotor from a jet engine. (*Source:* Reproduced by permission of American Welding Society, ©*Welding Handbook*).

electronics. Many applications involve the welding of small, intricate components due to the ability of this process to precisely locate and concentrate the heat. Some Electron Beam Welding equipment is capable of production welding, but since the equipment cost is extremely high, it is also common for manufacturers to utilize an Electron Beam Welding job shop to conduct their welding.

In Summary, the Electron Beam Welding advantages and limitations are as follows:

Advantages:

- HED process allows for low overall heat input, which produces narrow heat-affected zones and minimal degradation to properties, as well as reduced residual stress and distortion
- Ability to weld the greatest single-pass thicknesses of any fusion welding process
- No filler metals or shielding gas required
- Vacuum environment makes it an ideal process for welding titanium that is very susceptible to interstitial contamination from the atmosphere
- No problems welding highly reflective metals
- Depth of focus greater than Laser Beam Welding, which provides more flexibility regarding gun to work piece distance
- Beam can be easily manipulated with deflection coil
- Excellent weld "aesthetics"

Limitations:

- The need for a vacuum limits production speed and part size
- Very expensive equipment
- Machines are not portable
- Joints require very precise fit-up
- Fast cooling rates may create brittle microstructures with some steels
- Electron Beam Welding produces dangerous X-rays resulting in the requirement for shielding in the chamber

5.4 Test Your Knowledge

I) Fundamental Concepts—True/False

The following true/false questions pertain to some of the most important fundamental concepts in this chapter:

1) HED processes are known to deliver much more heat to the part being welded than arc welding processes.
2) HED processes produce fast cooling rates, which can be a problem when welding certain steels.
3) Laser Beam Welding produces greater depth of penetration than Electron Beam Welding.
4) A big advantage of Disk, Diode, and Fiber Lasers is they are much more electrically efficient than Nd:YAG or CO_2 Lasers.
5) The amperage of an electron beam is known to be very low, in the range of 1 ampor less.
6) Electron Beam Welding can be a very good choice when welding titanium.

II) Solve a Welding Engineering Problem

The following challenge represents a typical problem Welding Engineers might encounter in their career:

The high production company you work for currently has several CO_2 Lasers that were used for an older application but are no longer being used. Your boss sees a good opportunity to put these lasers into production if they can be mounted to robots, and she/he seeks your advice regarding whether this is possible. What is your answer and your explanation?

Recommended Reading for Further Information

ASM Handbook, Tenth Edition, Volume 6—"Welding, Brazing, and Soldering", ASM International, 1993.

AWS Welding Handbook, Ninth Edition, Volume 3—"Welding Processes, Part 2", American Welding Society, 2007.

Laser Material Processing, Third Edition, Springer-Verlag London Ltd, 2003.

6

Other Approaches to Welding and Joining

Whereas the previous chapters covered in some detail the conventional industrial metal welding process families of arc, resistance, solid-state, and HED welding, this chapter provides a brief overview of many other ways both metals and nonmetals are joined. The word "joining" is introduced here to include processes like brazing, soldering, and adhesive bonding, which are not true welding processes. New to this second edition is an introduction to additive manufacturing (AM) which is rapidly becoming an important topic in manufacturing that relies on several common welding processes. Methods for mechanical fastening are beyond the scope of this book, and therefore, are not covered.

6.1 Brazing and Soldering

Brazing and Soldering are joining processes that rely on filler metals that melt at temperatures below the melting temperature of the base metal being joined. Because the processes do not depend on base metal melting, they are not considered welding processes. Per AWS, the process is considered Brazing when the filler metal melts above 450°C (842°F), and considered Soldering when the filler metal melting temperatures are below 450°C. Although these processes do not exceed the melting point of the base metals being joined, it is not unusual for Brazing to produce some localized base metal melting due to a chemical reaction with the braze alloy which lowers the melting point of the base metal. But excessive melting can sometimes create a problem known as base metal erosion.

Brazing and Soldering both require that the molten filler metal wets and flows through the joint via capillary action. Capillary action relates to the force of attraction that can occur between a liquid and a solid. A classic approach to verifying capillary action is to dip a small diameter tube into a liquid and observe the extent to which the liquid rises up the tube (Figure 6.1). If there is good capillary action, the liquid will rise quite high into the tube, seemingly defying gravity. This same concept applies to joint gaps during brazing and soldering. If good capillary action is not achieved, the joint may not be completely filled. Many factors affect capillary action, including the joint gap width, uniformity of temperature throughout the joint, viscosity of the molten filler metal, and cleanliness of the surfaces.

One big advantage of processes that do not require base metal melting is that both metals and non-metals (i.e., ceramics and composites) can be joined. Brazing is also used to join metals to nonmetals. There are numerous Brazing and Soldering filler metal compositions available in various forms, including pastes and inserts. Fluxes are often used to enhance the capillary action through cleaning, deoxidizing, and modifying the surface tension between the liquid filler material and the material being joined. Since joint strength is directly related to the area of the joint, these

Welding Engineering: An Introduction, Second Edition. David H. Phillips.
© 2023 John Wiley & Sons, Inc. Published 2023 by John Wiley & Sons, Inc.
Companion Website: www.wiley.com/go/Phillips/WeldingEngineeringIntroduction

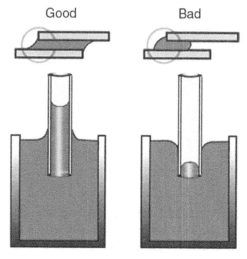

Figure 6.1 Approach to verifying capillary action.

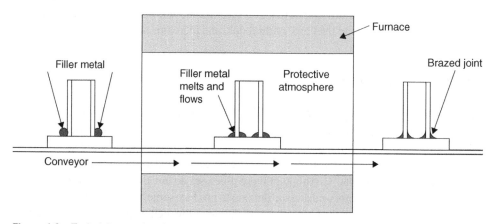

Figure 6.2 Typical furnace brazing operation. (*Source:* Reproduced by permission of American Welding Society, ©*Welding Handbook*).

processes are most conducive to lap joints that allow for the creation of large joint areas. There is a wide variety of equipment and approaches available for these processes, including furnaces, dip tanks (baths), induction coils, and torches. Figure 6.2 depicts a typical furnace brazing operation using a preplaced filler metal. The applications for Brazing and Soldering are diverse and are frequently the processes of choice when welding is not a viable option.

In summary, the advantages and limitations of Brazing and Soldering are as follows:

Advantages:

- Ability to join nonmetals and metals-to-non-metals
- Minimal base metal degradation and distortion
- Can be relatively economical when conducted in "batch" operations

Limitations:

- Processes are often slow and labor intensive
- Joint strength depends on size of joint area, so joint designs are limited

- Significant joint preparation usually required
- Equipment can be expensive
- Base metal erosion can be a problem with some Brazing operations
- Joints typically exhibit poor ductility
- Service temperature will be limited, especially with joints that are soldered

6.2   Welding of Plastics

Plastics (or polymers) represent a wide range of synthetic materials that offer a variety of benefits including light weight, corrosion resistance, and design flexibility. The word "polymer" is derived from two Greek words and refers to molecules that are made up of many (poly) repeating units (mers). For example, polyethylene (Figure 6.3) is made up of many repeating ethene (ethylene) mer groups resulting in very long molecules or chains with a high aspect ratio (length/diameter).

Polymers are divided into two major groups: thermosets and thermoplastics. Thermoplastics are further divided by their molecular structure into amorphous or semicrystalline materials (Figure 6.4). As the name implies, thermosets are thermally set to a final shape after they undergo an irreversible chemical reaction. Once formed, they cannot be remelted, and therefore, cannot be welded. Thermoplastics can be deformed plastically once they are reheated, and they solidify again when they are cooled. In the case of thermoplastics, interactions between molecules occur through secondary chemical bonds that can be broken at elevated temperatures causing melting. As a result, thermoplastics are easily welded with a variety of processes.

Figure 6.5 illustrates the difference in transition temperatures between the two types of thermoplastics. Amorphous thermoplastics are characterized by a randomly arranged molecular structure like wet spaghetti and with no defined melting point. When heated, they

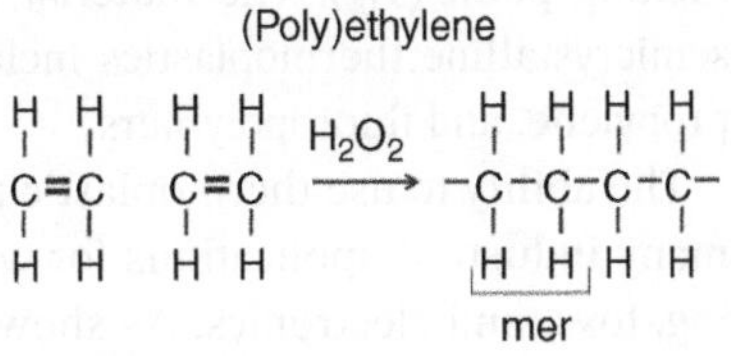

(Poly)ethylene

Figure 6.3   Polyethylene. (*Source:* Dr. Avi Benatar, The Ohio State University).

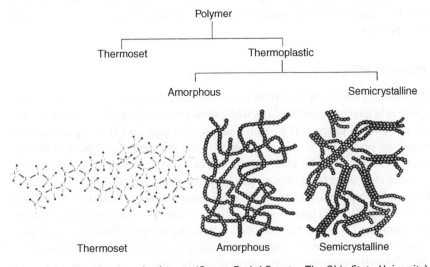

Figure 6.4   Classification of polymers. (*Source:* Dr. Avi Benatar, The Ohio State University).

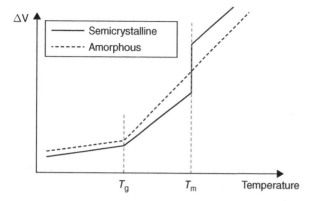

Figure 6.5 Semicrystalline and amorphous thermoplastics. (*Source:* Dr. Avi Benatar, The Ohio State University).

gradually soften as they pass from a rigid state, through a glass transition temperature (T_g), into a rubbery state, followed by a liquid flow in a true molten state. Upon cooling, solidification is equally gradual. Major types of amorphous thermoplastics include ABS, Styrene, Acrylic, PVC, and Polycarbonate.

Semicrystalline thermoplastics are characterized by a combination of amorphous regions and crystallites that have a very orderly and repeated molecular arrangement with a well-defined melting point (T_m). The material remains solid until it reaches the melt temperature. Major semicrystalline thermoplastics include acetal, nylon (polyamide), polyester, polyethylene, polypropylene, and fluoropolymers.

The ability to use thermoplastic polymers quite often requires that they be welded. There are many industrial applications for welding polymers, including automotive, appliances, packaging, toys, and electronics. As shown in Figure 6.6, there are numerous approaches to welding plastics. It is common to divide them into processes that introduce the heat for welding either externally or internally. External heating processes rely on conduction or convection to heat the weld surface. Mechanical internal heating methods rely on conversion of mechanical energy into heat, while electromagnetic internal heating methods rely on absorption and conversion of electromagnetic radiation into heat. The following is a brief review of some of the more common welding processes for plastics.

6.2.1 Hot Tool (Plate) Welding

Hot Tool Welding involves the use of a heated plate placed between the parts to be welded (Figure 6.7). In step I, the plate is heated between the parts while some pressure is applied. This is known as the matching phase, and its purpose is to eliminate any warpage and create good contact between the parts and the plate. Heat continues in step II with no pressure, allowing a melted region to form on each of the parts. The plate is removed in step III and the parts are quickly brought into contact with each other before the molten layers begin to solidify. In the final step IV, forging pressure is applied, which produces the weld as it squeezes molten material outward into what is referred to as the flash. Hot tool (plate) welding is a relatively slow welding process that has very wide operating windows. Therefore, the process is robust and reliable although it has a limited temperature range. It is used in a wide range of applications including some critical applications such as joining of polyethylene pipes for transport of natural gas.

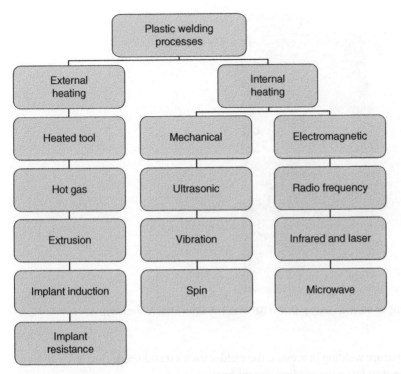

Figure 6.6 There are many methods for welding plastics. (*Source:* Dr. Avi Benatar, The Ohio State University).

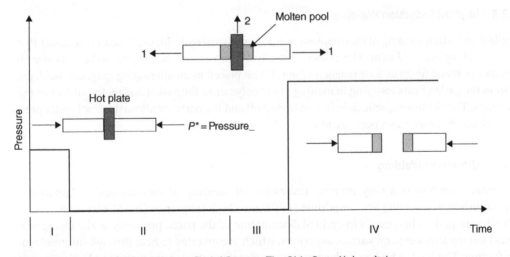

Figure 6.7 Hot tool welding. (*Source:* Dr. Avi Benatar, The Ohio State University).

6.2.2 Hot Gas Welding

Hot Gas Welding (Figure 6.8) is a manual welding process that consists of directing a heated gas (air or nitrogen) at a temperature of 150–425°C (350–800°F) onto the weld joint and filler rod to soften/melt them. Filler rods are made of plastic that matches the plastic being welded. The hot gas temperature is usually controlled by adjusting the air flow rate through the electric cartridge heater

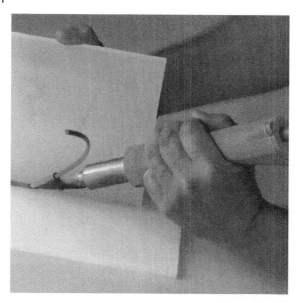

Figure 6.8 Hot gas welding of plastic. (*Source:* Dr. Avi Benatar, The Ohio State University).

in the weld gun. Much like arc welding processes, the welder uses a travel speed that is not so slow to overheat the parts or not so fast to create insufficient heating.

6.2.3 Implant Induction Welding

Implant induction welding of thermoplastics is accomplished by inductively heating a gasket that is placed along the weld joint. The gasket is usually a composite of the polymer to be welded with conductive metal fibers or ferromagnetic filler. When placed in an alternating magnetic field, the filler in the gasket heats resulting in melting of the polymer in the gasket and on the surface of the two parts. The electromagnetic field is then turned off and the parts are allowed to cool under pressure. The gasket becomes a permanent part of the assembly.

6.2.4 Ultrasonic Welding

Ultrasonic Welding is a very popular technique for welding of thermoplastics. Welding is accomplished by applying low amplitude (1–250 μm) high frequency (10–70 kHz) mechanical vibration to parts. This results in cyclical deformation of the parts, primarily at the faying surfaces (joining surfaces) and surface asperities, which is converted to heat through intermolecular friction. The heat, which is highest at the interface, is sufficient to melt the plastic creating the weld. Usually, to improve consistency, a triangular protrusion known as an energy director (Figure 6.9) is molded into one of the parts. The highest strain and greatest heating occur in the energy director, which melts and flows into the joint to create the weld. This process is similar to that used for metal welding except that the vibrations are perpendicular (instead of transverse) to the weld interface, and it relies on melting, which is not the case with metal Ultrasonic Welding.

6.2.5   Vibration Welding

Vibration Welding (Figure 6.10), which is a similar process to Linear Friction Welding (but also relies on melting), involves the rubbing of two parts together under pressure at a suitable frequency and amplitude to create frictional heat sufficient to melt the plastic and create the weld. As the figure indicates, the vibration can be in a linear motion or orbital motion. The amplitude of the vibrations typically ranges from about 0.5–2 mm, while the frequency of the vibrations ranges from 100 to 500 Hz. Vibration welding is fast, but the machines are expensive. While Ultrasonic Welding is limited to small parts, Vibration Welding can be performed on very large parts. The automotive and appliance industry sectors use Vibration Welding quite extensively. Automotive applications include front and rear light assemblies, fuel filler doors, spoilers, instrument panels, bumpers, power steering, and vacuum systems.

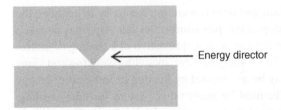

Energy director

Figure 6.9   An energy director is often used with ultrasonic welding.

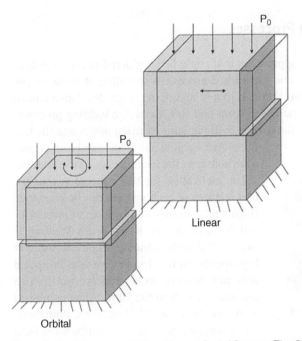

P_0

P_0

Linear

Orbital

Figure 6.10   Vibration welding. (*Source:* Dr. Avi Benatar, The Ohio State University).

6.3 Adhesive Bonding

Adhesive Bonding is a joining process for metals and nonmetals that uses an adhesive or glue, typically in the form of a liquid or a paste. It offers the major advantage of being able to join a wide array of materials but is limited by joint strength and applicable service conditions. Many of the fundamentals are similar to Brazing and Soldering, such as the need for wetting and capillary action, and lap joint designs that rely on large joint areas for strength. Adhesives are categorized as thermosetting or thermoplastic. Thermosetting or hot melt adhesives require a chemical reaction to cure and cannot be remelted once cured. They are the most common adhesives for structural applications. Thermoplastic adhesives soften when heated and harden when cooled, a process that can be continually repeated. They are generally not used for structural applications because of their poor mechanical properties (in particular, creep) and low temperature range service temperatures.

Surface preparation is usually a very important and time-consuming step prior to applying the adhesive, and typically includes cleaning and degreasing procedures. Further steps may be taken such as roughening the surface to increase the bond area, and application of a coating to protect the treated surface. Adhesives may be applied in a variety of ways, including caulking and spray guns, dipping, rollers, and brushes. Curing may be accelerated by heating or some other energy source. In addition to joining, adhesives may be used for many other reasons including sealing, electrical insulation, and sound suppression. Some of the larger users of Adhesive Bonding include the automotive, appliance, and electrical/electronics industry sectors.

6.4 Novel and Hybrid Welding Processes

Recent developments in welding include approaches that could be considered novel, as well as hybrid, or those that combine more than one established process. A sampling of these unique approaches is briefly reviewed here. A hybrid process that is being extensively developed known as Hybrid Laser Welding combines both the Laser Beam and Gas Metal Arc Welding processes (Figure 6.11). This approach uses a head that carries both the laser focusing optics and the Gas Metal Arc Welding gun. The laser beam creates a keyhole near the leading edge of the puddle.

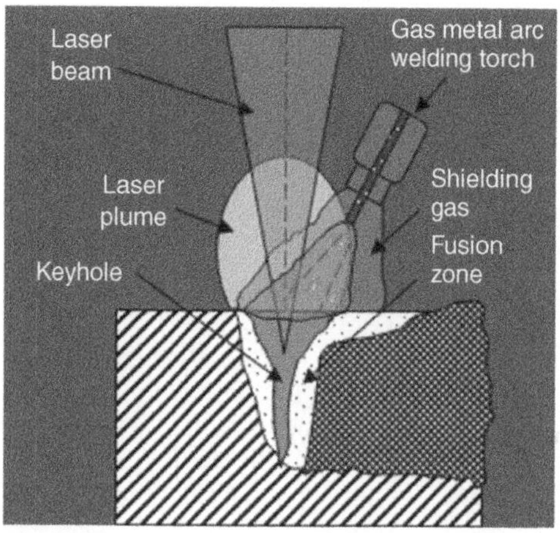

The motivation for this concept is that the best features of each process can be combined (Figure 6.12) to create an even better process for certain applications. Laser Beam Welding is capable of deep penetrating single pass welds at high speeds but requires precise joint fit-up and does not produce weld reinforcement which can add strength to the joint. When combined with Gas Metal Arc Welding, these limitations are eliminated, and the best attributes of both processes are realized. The laser beam also helps stabilize the arc, which can be of particular benefit when welding titanium.

Plasma and Gas Metal Arc Welding have also been combined into a hybrid process. The

Figure 6.11 Hybrid laser welding. (*Source:* Image provided by the Lincoln Electric Company, Cleveland, OH, USA).

benefits are similar to those of the Laser Hybrid Welding approach, but the equipment has the potential to be much less expensive. The combination of Friction Stir Welding with ultrasonic energy is being explored as an approach to producing Friction Stir Welds with greatly reduced forces, and therefore, smaller machines.

A novel arc welding process with the trade name of "TiPTiG" (Figure 6.13) combines Gas Tungsten Arc Welding with a continuously fed wire. This provides a process that offers the precise heat control Gas Tungsten Arc Welding is known for with much faster welding speeds. A unique approach to resistance welding with the trade name of "DeltaSpot" (Figure 6.14) is a Resistance Spot Welding process that uses a continuously fed metal tape between the electrodes and the part being welded. A major benefit is significant reduction in electrode wear and greater weld consistency, especially when welding difficult-to-weld metals like aluminum. The type of metal tape can be varied as well. Metal tapes of greater electrical resistance can be selected to generate more heat when welding highly conductive metals.

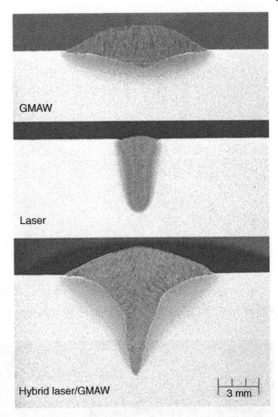

GMAW

Laser

Hybrid laser/GMAW

3 mm

Figure 6.12 Hybrid laser welding combines the best features of laser beam and gas metal arc welding. (*Source:* Edison Welding Institute).

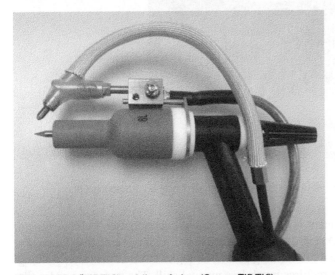

Figure 6.13 "TiP TiG" welding of pipe. (*Source:* TiP TiG).

Figure 6.14 "DeltaSpot" welding. (*Source:* Fronius International GmbH).

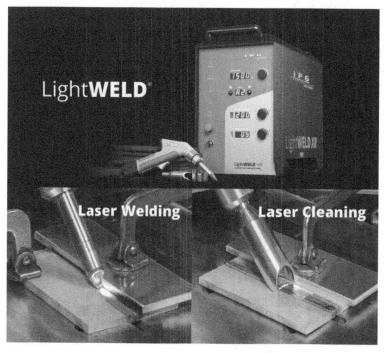

Figure 6.15 "LightWELD" handheld laser welding system. (Image courtesy of IPG Photonics Co).

A recent very interesting development is a portable handheld manual Laser Welding system (Figure 6.15) known as LightWELD (LightWELD is a registered trademark of IPG Photonics Corporation). As compared to manual arc welding processes, LightWELD offers faster welding speeds, much lower heat input, and requires much less manual skill. The system provides the option of using a continuously fed filler wire, or no filler at all. A wobble welding feature allows for wider weld beads to be made if needed. The laser can also be used to preclean the joint prior to welding and postweld to improve visual finishes.

In summary, these are just a few of the unique approaches to welding that are being explored and developed. They all offer interesting benefits, but usually at the expense of additional process complexity, greater cost, and an unproven track record in a manufacturing environment. As with any welding process, these approaches will likely find their "niche" markets, where added equipment cost and complexity may be justified by the cost savings and other benefits realized in manufacturing.

6.5 Additive Manufacturing

Additive Manufacturing (AM) is a rapidly growing manufacturing concept that has the potential to revolutionize the way many products are developed and manufacturing over the next 10–20 years. Additive Manufacturing for metal parts can be simply described as 3-D printing of metals. A major benefit of AM is the significant reduction in the time associated with designing and building a part. AM begins when a part drawing on a digital CAD file is loaded into the AM machine, which in turn, converts the drawing to individual thin layers. The machine then builds the physical part one layer at a time, using a variety of welding processes (there are also other approaches that don't rely on welding). There may or may not be some minor machining required after the part as built, but the result is a part of complex geometry can be built very quickly and at low cost. There are many welding processes associated with Additive Manufacturing, but this section will focus primarily on the very common laser beam approach. As discussed previously, Laser Beam Welding is a high energy density process which provides for fast welding speeds and minimal heat input, attributes that make it very attractive for Additive Manufacturing of metals. When utilizing a laser, there are two common AM methods for building the part known as powder bed fusion and direct energy deposition.

With powder bed fusion (Figure 6.16), thin layers of metal powder are raked onto a flat surface one thin layer at a time. With each layer, the laser fuses the metal following the pattern of the CAD file slice, and this process is repeated one layer at a time until the part is built. Upon completion, the part is removed, and the unfused powder is left behind. This approach offers the advantage of producing extremely intricate parts, but the process is slow and part sizes are limited.

Figure 6.16 Laser powder bed fusion additive manufacturing.

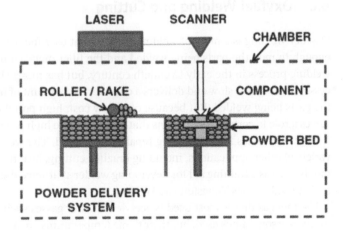

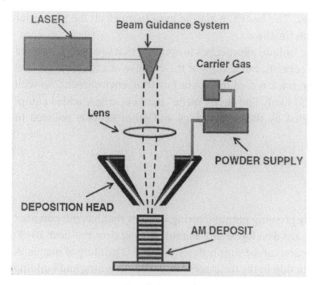

Figure 6.17 Laser direct energy deposition additive manufacturing.

Direct energy deposition (Figure 6.17) delivers the metal to be used for the build directly into the beam, and as with the powder bed, the part is built one layer at a time. The example shows metal powder being fed into the beam, but solid wire can be used as well. Direct energy deposition can be faster than powder bed fusion, but at the sacrifice of reduced ability for building complex geometries.

Other approaches to AM of metals that utilize fusion welding processes include electron beam (powder bed fusion and direct energy deposition) and arc (direct energy deposition). Another slightly different approach involves building the part with layers of foil using the solid-state welding process Ultrasonic Welding. All the AM approaches offer certain advantages and disadvantages. Therefore, the best choice for a given application will be driven by a variety of factors, including part size and complexity, required build speed, material type, and the extent of any post build processing needed such as machining.

6.6 Oxyfuel Welding and Cutting

Oxyfuel Welding is a manual welding process that uses fuel gas and oxygen to create a flame hot enough to melt the parts being welded and the filler metal (if used). It was a common industrial welding process in the early twentieth century, but has mostly been replaced by arc welding processes because it is slow and delivers a considerable amount of heat in an oxidizing atmosphere to the parts being welded. But because of its low cost, high portability, and process flexibility it still serves a role in certain applications that do not require high production rates. Its portability makes it an ideal process for performing repairs in the field. The welding equipment can be used for a variety of other applications, including brazing, cutting, heat treating, and bending. It is also commonly used as a training aid for developing welder skill since the welding technique is very similar to that used for Gas Tungsten Arc Welding.

The fuel gas that is most used is acetylene. Other gasses such as propane, natural gas, and propylene are sometimes used, but their flame temperatures are not as high as the acetylene flame. As illustrated in Figure 6.18, flame temperatures when acetylene is used as the fuel gas can approach 3300°C (~6000°F). The flame profile consists of an inner cone, an acetylene feather, and an outer envelope, with the highest temperatures located near the end of the acetylene feather. The oxygen-acetylene ratio can be adjusted to produce various flame types, and Figure 6.19 shows the steps required to achieve the most common flame, the neutral flame. As the oxygen content is increased, the flame type progresses from an orange sooty flame to a carburizing flame to a neutral flame. Adding more oxygen produces an oxidizing flame. Carburizing flames produce less heat and oxidation but can actually increase the carbon content while welding steel which can reduce ductility. Oxidizing flames are very hot, which is good for cutting or welding metals of high thermal

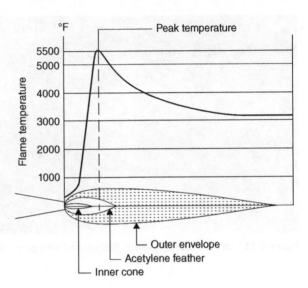

Figure 6.18 Oxyacetylene flame temperatures. (*Source: Welding Essentials, Second Edition*).

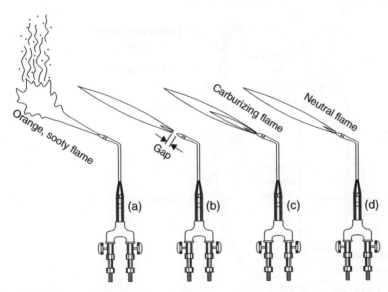

Figure 6.19 Steps to achieve a neutral flame—(a) ignite the flame with acetylene valve slightly open, (b) continue to open acetylene valve until smoke disappears and a gap is seen, (c) slowly open oxygen valve to produce white cone, and (d) continue to open oxygen valve to produce a smaller, clearly defined cone. (*Source: Welding Essentials, Second Edition*).

conductivity such as alloys of copper. However, oxidizing flames should not be used for welding steel because of the extreme amount of oxidation that will occur.

Oxyfuel Cutting (Figure 6.20), which is commonly used for cutting steel, uses the same equipment (Figure 6.21) as welding, but features a cutting torch instead of a welding torch. It can be a manual or mechanized process. The cutting torch heats the steel to its kindling temperature of approximately 870°C (1600°F). While the role of the flame is to preheat the steel, the actual cutting is conducted by a stream of pure oxygen delivered through the torch which burns (oxidizes) the steel to create the cut and clear the molten metal away. The oxidation process generates additional

Figure 6.20 Oxyfuel cutting. (*Source:* Reproduced with permission of American Welding Society).

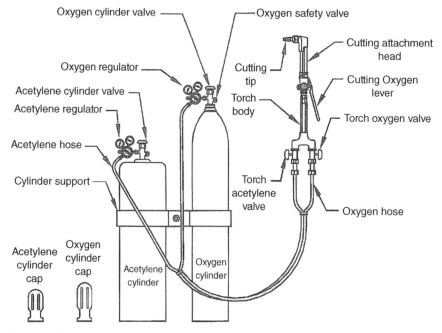

Figure 6.21 Oxyacetylene cutting equipment. (*Source: Welding Essentials*, Second Edition).

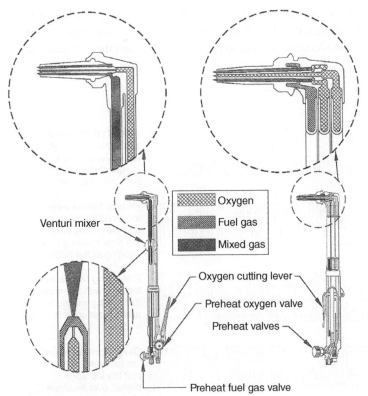

Figure 6.22 Common cutting torches—injector type on the left, and mixing chamber on the right. (*Source: Welding Essentials*, Second Edition).

heating, which allows the cutting to continue at relatively fast speeds. Because the process relies on the oxidation reaction with steel, its use is restricted mainly to ferrous metals. The width of the cut is known as the kerf. The same gasses used for welding can be used for cutting, but again, acetylene is the most common because it produces the most heat.

Figure 6.22 shows the two common types of cutting torches, the injector torch and mixing chamber (or positive pressure) torch. The mixing chamber torches are used when there is sufficient acetylene pressure available for mixing in the torch. When there is not sufficient acetylene pressure, the injector torch is used since it incorporates a venturi mixer that draws the fuel into the torch via venturi action. A third option is a cutting accessory head (Figure 6.21), which provides for easy and quick changeover from a welding head. Although the cutting accessory head is convenient, it is limited in terms of cutting speeds and the thicknesses of steel that can be cut.

A good quality cut consists of a combination of proper kerf and drag (Figure 6.23).

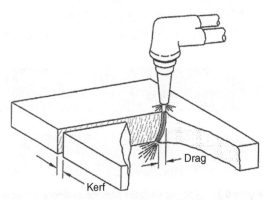

Figure 6.23 Kerf and drag. (*Source: Welding Essentials*, Second Edition).

When attempting to cut too fast, excessive drag will eventually result in the loss of cutting action. Reverse drag, which can occur if oxygen flow is excessive, can result in rough cut edges. The cutting surface features can provide evidence of cut quality, and reasons for less than an ideal cut as illustrated on Figure 6.24.

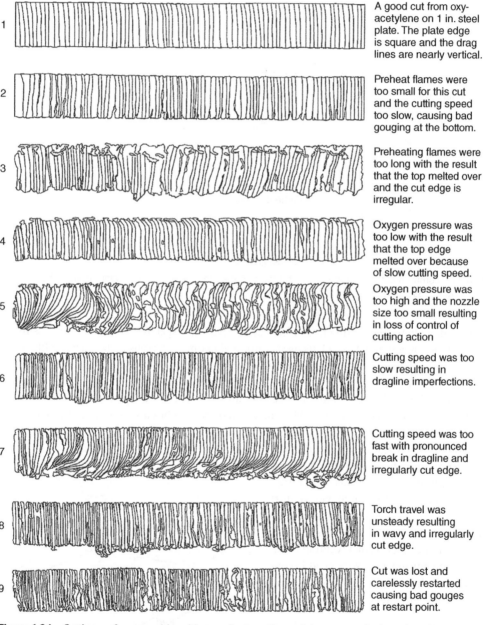

1 — A good cut from oxy-acetylene on 1 in. steel plate. The plate edge is square and the drag lines are nearly vertical.

2 — Preheat flames were too small for this cut and the cutting speed too slow, causing bad gouging at the bottom.

3 — Preheating flames were too long with the result that the top melted over and the cut edge is irregular.

4 — Oxygen pressure was too low with the result that the top edge melted over because of slow cutting speed.

5 — Oxygen pressure was too high and the nozzle size too small resulting in loss of control of cutting action

6 — Cutting speed was too slow resulting in dragline imperfections.

7 — Cutting speed was too fast with pronounced break in dragline and irregularly cut edge.

8 — Torch travel was unsteady resulting in wavy and irregularly cut edge.

9 — Cut was lost and carelessly restarted causing bad gouges at restart point.

Figure 6.24 Cutting surfaces provide evidence of cut quality and the reasons for less-than-ideal cutting action. (*Source: Welding Essentials*, Second Edition).

6.7   Other Cutting Processes

6.7.1   Plasma Cutting

Compared to Oxyfuel Cutting, Plasma Cutting (Figure 6.25) offers faster cutting speeds, higher quality cuts, and smaller heat-affected zones. It can also be used for both manual and mechanized cutting. Plasma cutting does not rely on oxidizing the metal, but instead uses a concentrated arc plasma and high velocity flow of gas to cut and eject the molten metal. As a result, it can be used for both ferrous and nonferrous metals. Arc currents can be as high as 1000 amps and as low as a few amperes. Because of arc temperatures that are in the range of 14,000°C (25,000°F), it can cut metals with very high melting temperatures.

Figure 6.25   Plasma cutting. (*Source: Reproduced by permission of American Welding Society, ©Welding Handbook*).

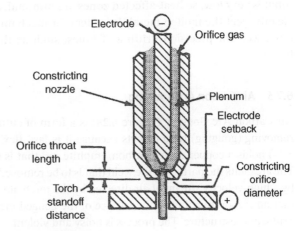

Figure 6.26   Mechanized plasma cutting system. (*Source:* Lincoln Electric Company).

Plasma cutting generates a tremendous amount of both noise and fumes. To reduce both, water tables are often used. In this case, cutting may be above or below the water, and sometimes additional water flow is supplied through the torch. The water in the table and the torch traps the fumes and suppresses the noise. Mechanized cutting systems (Figure 6.26) that may or may not rely on water are common because they can rapidly produce very straight, consistent cuts.

6.7.2 Laser Beam Cutting

Laser Beam Cutting (Figure 6.27) provides for even faster cutting of greater quality than Plasma Arc Cutting, but with the limitation of higher equipment cost. Laser Beam Cutting kerfs can be extremely narrow, and very thin sheets can be cut with great precision. Heat input is very low, so heat-affected zones are minimal. Assist gas may or may not be needed to cleanly eject the molten metal. It is ideal for mechanized and automated cutting, and works quite well for precision drilling of holes, such as the cooling holes in gas turbine engine blades.

6.7.3 Air Carbon Arc Gouging

Air Carbon Arc Gouging (Figure 6.28) is a form of cutting that is a very common approach for removing (gouging) weld defects because it is fast, flexible, and inexpensive. The torch (Figure 6.29) holds a copper-coated carbon/graphite rod that is slowly consumed by the heat of the arc. The arc melts the portion of the weld that is to be removed, and the molten metal is rapidly rejected by a high-velocity stream of air directed by the torch along the rod. Care must be taken to ensure that no carbon remains on the surface of the gouged metal because it could result in a hard and brittle microstructure. The process is noisy and violent.

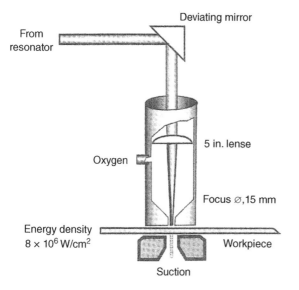

Figure 6.27 Laser beam cutting with oxygen assist gas. (*Source: Welding Essentials*, Second Edition).

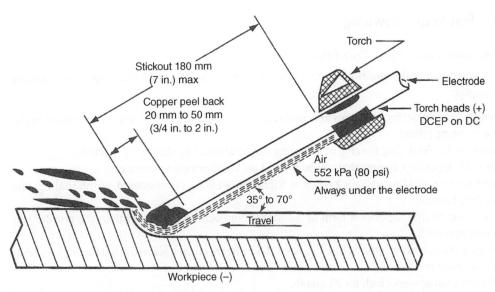

Figure 6.28 Air carbon arc gouging. (*Source:* Reproduced by permission of American Welding Society, ©*Welding Handbook*).

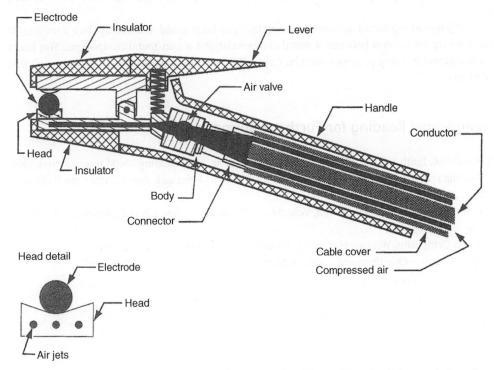

Figure 6.29 Typical air carbon arc cutting torch cross-section. (*Source:* Reproduced by permission of American Welding Society, ©*Welding Handbook*).

6.8 Test Your Knowledge

I) Fundamental Concepts—True/False

The following true/false questions pertain to some of the most important fundamental concepts in this chapter:

1) Brazing and Soldering processes are conducted at temperatures below the melting point of the metals being joined.
2) Capillary action during brazing or soldering can be affected by the joint gap width.
3) The only types of polymers that can be welded are the thermoset polymers.
4) A process for welding plastics known as Vibration Welding is considered an internal heating welding process.
5) The direct energy deposition form of Additive Manufacturing is slower than the powder bed fusion approach.
6) Of the various cutting approaches discussed, Laser Beam cutting is considered the highest quality and most precise.
7) Oxyfuel cutting works well for all metals.

II) Solve a Welding Engineering Problem

The following challenge represents a typical problem Welding Engineers might encounter in their career:

On your job working for an automotive company, you been asked to come up with a method to create a strong connection between a metal component and a non-metal component. You know that a traditional welding process would be impossible, but you have an idea of an approach that should work.

Recommended Reading for Further Information

ASM Handbook, Tenth Edition, Volume 6—"Welding, Brazing, and Soldering", ASM International, 1993.
AWS Welding Handbook, Ninth Edition, Volume 2—"Welding Processes, Part 1", American Welding Society, 2004.
AWS Welding Handbook, Ninth Edition, Volume 2—"Welding Processes, Part 2", American Welding Society, 2007.
Plastics and Composites Welding Handbook, Hanser Gardner Publications, Inc., 2003.
Welding Essentials—Questions and Answers, Second Edition, International Press, 2007.

7

Design Considerations for Welding

7.1 Introduction to Welding Design

In Welding Engineering, the concept of welding design encompasses a broad range of subjects that includes weld and joint types, economics, weld sizing for various loading conditions, welding symbols, computational modeling, mechanical properties and testing, residual stress and distortion, and heat flow. Welding codes often play an important role in welding design, as does knowledge of the material being welded, how it was processed, and what welding filler metal is being used. Welding design often depends highly on a thorough understanding of the expected loading conditions and desired service life of the weldment. For example, a weldment that is going to undergo fatigue conditions will have different design criteria than one that is subject to pure tensile or compressive loading.

Distortion and residual stress that results from welding are often important considerations for design since significant amounts of either one may affect whether the weldment is acceptable. Finally, the Welding Engineer must be knowledgeable of the many methods for nondestructively inspecting the weld before it enters service. Most of the welding design concepts covered in this chapter focus on arc welding, but it is important to remember that similar considerations are associated with all welding processes.

Many of these concepts require knowledge of the basic mechanical and physical properties of metals. For example, a metal with a higher coefficient of thermal expansion (COE) will distort more when welded than one with a lower coefficient. Another example is the significant loss in tensile strength in the heat-affected zone some metals will undergo when welded. Therefore, it is useful to first briefly review the mechanical and physical properties of metals that are of greatest importance to the Welding Engineer. Mechanical properties, such as tensile strength, are those that characterize the way a metal performs under various mechanical loading conditions, while physical properties are dictated by the chemistry of the metal and include properties such as melting point and electrical conductivity.

7.2 Mechanical Properties

7.2.1 Yield Strength

Yield strength is a mechanical property that can be derived from the stress-strain curve of a tensile test and represents a stress level at which a metal deviates a specified amount from the elastic

Welding Engineering: An Introduction, Second Edition. David H. Phillips.
© 2023 John Wiley & Sons, Inc. Published 2023 by John Wiley & Sons, Inc.
Companion Website: www.wiley.com/go/Phillips/WeldingEngineeringIntroduction

portion of the curve to the plastic portion. More simply, when a metal bar is deflected (strained) by a load and returns to its original shape after the load is released, the strain exhibited is said to be elastic. If the deflection has caused permanent deformation, it has produced a plastic strain, which means the yield strength of the metal has been exceeded. Since with ductile metals exceeding the yield strength causes plastic deformation which typically would be considered a structural failure, yield strength is often used for structural fabrication design criterion. Therefore, a common approach to classifying structural steels is by their yield strength. For example, A36 is an ASTM specification for a family of structural steels that exhibit a minimum yield strength of 36 ksi.

7.2.2 Tensile Strength

Tensile strength (or ultimate strength) is another mechanical property that can be determined from a stress-strain curve and represents the maximum stress on the curve. In simple terms, tensile strength is a measure of the maximum load a material can support. Ductile metals will continue to stretch (plastically deform) after the maximum tensile strength has been reached, and therefore, the maximum tensile strength does not actually represent the amount of strain the metal can tolerate before fracture. Because of the significant plastic deformation that can occur with ductile metals prior to fracture, tensile strength is not normally used for design purposes (yield strength is). But since it is easy to identify tensile strength from a stress-strain curve, it is often used as an approach to quality monitoring or for comparing different materials. With brittle metals that exhibit little to no plastic deformation when tested, the maximum tensile strength and strength at fracture are the same, so it is more common to use tensile strength as a design criterion with such metals.

7.2.3 Ductility

Ductility refers to the amount of plastic deformation that a metal undergoes prior to fracture. It can be determined with a tensile test either from the percent elongation or percent reduction of area of the test sample after it is fractured. Metals with poor ductility do not perform well when subjected to impact loads. When welding certain metals such as steels, the heat-affected zone may exhibit significant reductions in ductility which will degrade the impact properties of the joint.

7.2.4 Fatigue Strength

Fatigue strength relates to a metal's behavior when subjected to cyclic loading conditions. It is often defined as the maximum stress range that can be sustained for a stated number of cycles without failure. Fatigue properties are greatly affected by sharp geometrical features in a part or fabrication that create regions of stress concentration during loading. This is because such localized regions can experience stresses exceeding the yield stress of the material, even though the loading conditions would not be expected to exceed the yield stress of the part. Welded joints quite often produce regions where stresses may concentrate such as the toe of the weld (Figure 7.1), and therefore, represent favorable sites for fatigue cracking.

7.2.5 Toughness

Toughness refers to the ability of a material to resist fracture and absorb energy under impact-type loading. Toughness tests such as the Charpy V-Notch test (Figure 7.2) involve machining a sharp notch in the test specimen and impacting it from the back side causing it

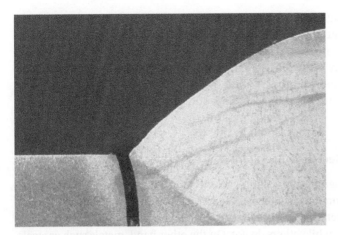

Figure 7.1 Typical location of a fatigue crack.

Figure 7.2 Impact test machine for assessing toughness.

to break at the notch. Good toughness requires a combination of ductility and tensile strength, so it is possible for a material to have good ductility but poor toughness if its tensile strength is low. At the same time, a material with high tensile strength will perform poorly under impact loading if its ductility is low. Brittle failures due to poor toughness can be sudden and catastrophic. Also, it's important to note that there is a difference between Charpy V-Notch toughness and "true" fracture toughness. These subjects are covered in more detail in Chapter 13.

7.2.6 Mechanical Properties—Effect of Temperature

The temperature of the metal can significantly degrade its mechanical properties, and hence the service conditions become an important consideration. Higher temperatures will reduce both the tensile and yield strength. Some metals such as nickel-based alloys maintain good mechanical properties to very high temperatures while others do not. On the other hand, metals such as steels may exhibit a sharp reduction in ductility at low temperatures. This is known as the ductile-to-brittle transition temperature and can play an important role in the selection of steels and the applicable service conditions.

7.3 Physical Properties

7.3.1 Thermal Conductivity

A metal with higher thermal conductivity will transfer heat more rapidly than one with lower conductivity. During welding, the thermal conductivity of the metal may affect a variety of considerations. For example, during resistance welding, a metal with poor thermal conductivity will heat up much faster than one with high thermal conductivity. During arc welding, the high thermal conductivity of copper makes it almost impossible to weld without preheat (unless the copper is very thin) since the heat is transferred away from the weld area so rapidly. To be more accurate, the rate at which heat will flow through a given metal is actually a function of its thermal diffusivity and is affected by the density and specific heat capacity (higher levels of each property will reduce rate of heat transfer) of the metal. But with metals of importance to welding, it is common to ignore these other properties since thermal conductivity alone usually sufficiently predicts the relative rate of heat flow.

7.3.2 Melting Temperature

Metals go from solid to liquid at a specific temperature known as its melting temperature. To the aspiring Welding Engineer, it might seem obvious that a metal with a lower melting point will require much less energy to weld than one with a higher melting temperature. This would be true if all metals had the same thermal conductivity (or diffusivity). As it turns out, the thermal conductivity of commonly welded metals plays a much bigger role than the melting temperature. For example, the melting temperature of aluminum is less than half that of steel, but requires greater amounts of heating when welding due to much higher rates of heat transfer out of the weld region. However, melting temperature does become important if the weld being attempted is a dissimilar metal weld involving metals with significant differences in their melting points.

7.3.3 Coefficient of Thermal Expansion

When a metal is heated it expands and when it is cooled it contracts. The amount of expansion and subsequent contraction is a function of the metal's physical property known as the coefficient of thermal expansion. Welding typically results in nonuniform heating and cooling. This in turn results in nonuniform expansion and contraction leading to residual stress and distortion (Figure 7.3) in weldments. As a result, metals with higher coefficients of thermal expansion such as austenitic stainless steels can be expected to distort more than those with lower coefficients such as plain carbon steels.

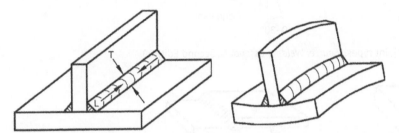

Figure 7.3 Weld distortion in fillet welds—the "T" and "L" refer to transverse and longitudinal residual stress in the weld region which creates distortion. (*Source: Welding Essentials*, Second Edition).

7.3.4 Electrical Conductivity

Electrical conductivity, which is inversely related to resistivity, is a property that affects how easily a material passes electrical current. It mainly plays a role with resistance welding processes because it directly affects how easily a material can be heated through I^2Rt (Joule) heating. Materials such as aluminum have much higher electrical conductivity than steels, and therefore, are much more difficult to resistance weld. Materials with higher electrical conductivity also exhibit higher thermal conductivity, which adds to the difficulty of generating sufficient Joule heating.

7.4 Design Elements for Welded Connections

Tubular and nontubular structural members are joined together by welded connections, which consist of a joint type and a weld type. Joint type refers to how the two work pieces being welded are oriented relative to each other, and weld type refers to the way the weld is placed in the joint. There are two main weld types, fillet and groove, but there are other approaches such as slot, seam, and spot. As the name implies, groove welds are those that are placed in grooves, and there are many versions of groove welds. Fillet welds are those that are placed in a corner formed by two mating work pieces. There are five basic joint types: butt, T, lap, corner, and edge.

7.4.1 Joint and Weld Types

Figure 7.4 shows the five basic joint types, while Figure 7.5 shows each of the joint types with a typical weld configuration. Each of the joints may be welded with a wide variety of weld types. Figures 7.6 and 7.7 show the many versions of groove welds, while fillet welds are shown in Figure 7.8.

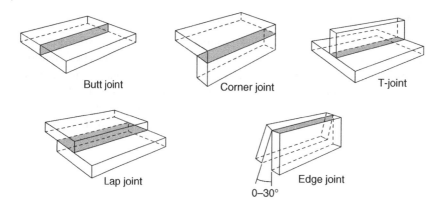

Figure 7.4 The five basic joint types. (*Source: Welding Essentials*, Second Edition).

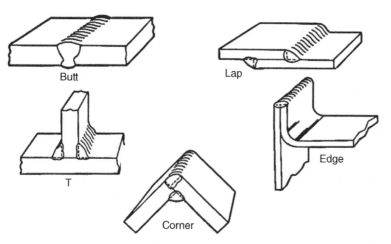

Figure 7.5 Typical approaches to welding the five joint types.

(a)

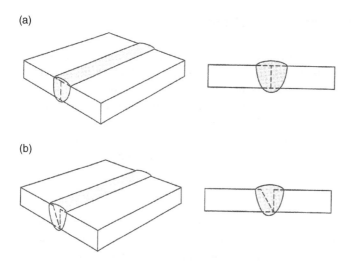

(b)

Figure 7.6 Examples of single-sided groove welds in butt joints. (a) Single-square-groove weld, (b) single-bevel-groove weld.

(Continued)

(c)

(d)

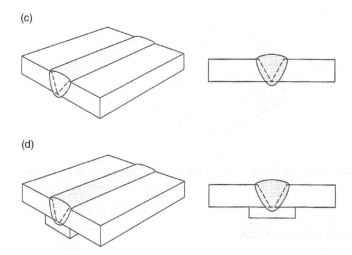

Figure 7.6 (Cont'd) (c) single-V-groove weld, and (d) single-V-groove weld with backing. (*Source:* Reproduced by permission of American Welding Society, ©*Welding Handbook*).

(a)

(b)

(c)

(d)

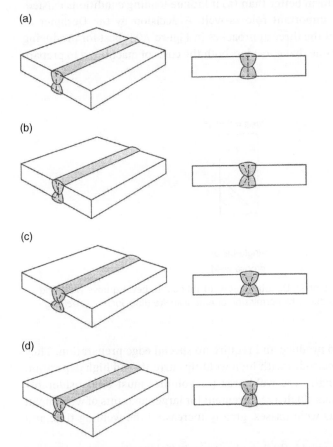

Figure 7.7 Examples double-sided groove welds in butt joints. (a) Double-square-groove weld, (b) double-bevel-groove weld, (c) double-V-groove weld, and (d) double-J-groove weld. (*Source:* Reproduced by permission of American Welding Society, ©*Welding Handbook*).

(a)   (b)

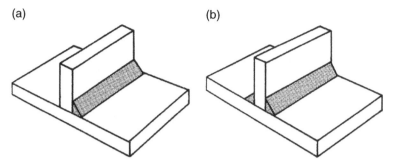

Figure 7.8 (a) Single and (b) double fillet welds in a T joint. (*Source:* Reproduced by permission of American Welding Society, ©*Welding Handbook*).

7.4.2 Joint and Weld Type Selection Considerations

Many factors affect the selection of the type of joint and weld to be used for a given application. A major priority of course is that the joint must be accessible to the welder. The choice of weld type may depend on the expected loading conditions. For example, in Figure 7.9 welds (b) and (c) would be expected to perform better than (a) if fatigue loading conditions existed. However, economics often play an important role as well. A decision by the Designer or Welding Engineer regarding which of the three approaches in Figure 7.9 is best for producing a T joint may simply be an economic one that considers both the costs of machining to prepare the joint and welding.

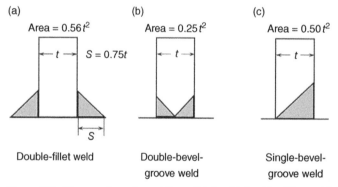

Figure 7.9 Three options for welding a T joint. (a) Double-fillet weld, (b) double-bevel-groove weld, and (c) single-bevel-groove weld. (*Source:* Reproduced by permission of American Welding Society, ©*Welding Handbook*).

Double-fillet welds, (a), are easy to produce and require no special edge preparation. They can be made using large diameter electrodes with high welding currents and high deposition rates, but to achieve sufficient strength, the weld area (or volume) must be considerably larger than in the other two approaches. Such a requirement for larger amounts of weld metal may result in the need for numerous weld passes, greatly increasing the welding time and welding cost.

By contrast, the double-bevel groove weld, (b), has about one half the cross-sectional area of the fillet welds. However, it requires costly edge preparation and the use of small diameter electrodes to make the initial root pass. In the third option, (c), a single-bevel groove weld requires about the same amount of weld metal as the double-fillet weld of (a), and thus, appears to have no apparent economic advantage. Compared to (a), the disadvantages of the single-bevel groove weld are that it requires edge preparation and a low-deposition root pass. From a design standpoint, however, it does offer direct transfer of force through the joint (as does (b)), so it would be expected to perform better than fillet welds when subjected to cyclic loading conditions.

Assuming loading conditions do not influence the choice of weld type, economic decisions regarding weld type selection may be dictated by the thickness of the parts being welded. As shown in Figure 7.10, the double-fillet weld approach may be the least expensive if the part thicknesses are less than 1in. But as part thicknesses increase, the amount of weld metal required rapidly increases, resulting in the potential need for a high number of weld passes. In extremely thick parts, 30–50 passes or more may be needed to fill a joint, which can be extremely time consuming. On the other hand, the double-bevel groove weld may be the most expensive when welding thinner parts because the edge preparation cost dominates. However, as part thickness increases, the associated increase in weld metal required, and therefore, the number of passes required, is much less than with the double-fillet weld. As a result, the considerable cost savings due to the greatly reduced number of weld passes required begins to dominate the edge preparation costs.

Figure 7.10   Effect of weld type on cost as a function of part thickness. (*Source:* Reproduced by permission of American Welding Society, ©*Welding Handbook*).

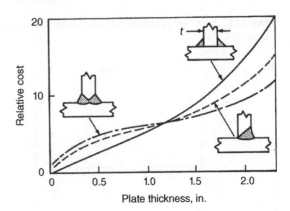

In addition to economics, other factors such as joint accessibility may play an important role in joint design. A larger groove angle (Figure 7.11) may be needed to allow easier access, especially for processes such as GMAW that use a bulky welding gun. As the figure also indicates, smaller groove openings (which save on the amount of weld metal required to fill the joint) will need larger root openings to achieve proper fusion. However, larger root openings are more likely to require backing bars.

In cases where backing bars are not possible (Figure 7.12), the proper root gap will allow sufficient fusion without producing excessive melt-through. Although not shown in the figure, a root face can make it easier to achieve proper root fusion without melt-through, but this adds additional cost to producing the joint. These are just some of the common considerations that go into choosing a joint design and weld type. As will be discussed later, applicable welding codes often make the decision process easier by providing prequalified joint designs and weld procedures.

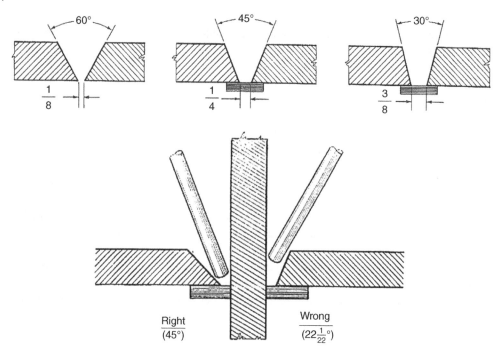

Figure 7.11   Joint design considerations. (*Source: Design of Weldments*, The James F. Lincoln Arc Welding Foundation).

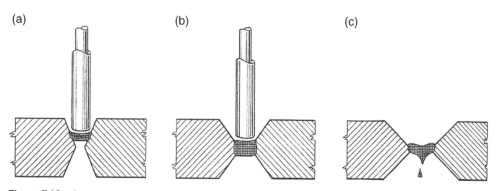

Figure 7.12   A proper root gap is critical (a) gap too narrow, (b) correct gap, and (c) gap too wide. (*Source: Design of Weldments*, The James F. Lincoln Arc Welding Foundation).

7.4.3   Weld Joint Nomenclature—Groove Welds

The development of a weld procedure requires knowledge of the nomenclature used to describe the various features of a weld joint. Figure 7.13 reveals the nomenclature used for groove welds, and includes terminologies used in both joint preparation prior to producing the weld, as well as that used to describe various features of the completed weld. Full penetration groove welds are required to achieve maximum strength, but in some cases, partial penetration welds are acceptable. Reinforcement refers to the measurable amount of weld buildup beyond the surfaces of the parts being welded and only applies to groove welds. Codes will often provide a maximum allowable amount of reinforcement. The weld toe is where the weld metal meets the base metal. This is often a region where problems such as fatigue cracking and undercut can occur.

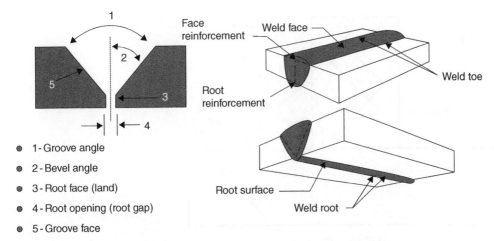

- 1 - Groove angle
- 2 - Bevel angle
- 3 - Root face (land)
- 4 - Root opening (root gap)
- 5 - Groove face

Figure 7.13 Groove weld nomenclature.

7.4.4 Weld Joint Nomenclature—Fillet Welds

Fillet weld nomenclature changes depending on whether the weld surface is convex or concave. To understand fillet weld nomenclature, it is helpful to draw the largest imaginary triangle (dotted lines in Figure 7.14) that can fit in the weld profile, with its height and base defined by the part geometry. With the convex fillet weld, the triangle helps define the theoretical and actual throat. The theoretical throat is the distance between the triangle hypotenuse and the corner of the triangle, while the effective throat adds an additional dimension that includes depth of fusion into the joint. The actual throat is the effective throat plus the convexity of the weld.

Weld leg and size are the same for the convex weld and are equal to the base and height of the imaginary triangle. With a concave fillet weld, actual and effective throat are the same measurements, but leg and size differ. Weld size and effective throat measurements are often used for quality control because both reflect the load-bearing capability of the joint. Also, weld sizes can easily be measured with simple tools. The reason the deviations between the concave and convex welds exist is to ensure conservatism when determining weld size or the effective weld throat.

7.4.5 Welding Positions

Welding position refers to the position of the welder relative to the location of the weld joint. Since welding is much easier in some positions than others, position becomes a very important aspect of qualifying both a welding procedure and a welder. For example, a welder may be skilled enough to qualify for a weld in a flat position, but not skilled enough to produce the same weld in a vertical position.

A welding position is designated by a number–letter combination. The number refers to the position, while the letter refers to the weld type: F for fillet weld or G for groove weld. When referring to a welding position, plate or pipe must be specified as well. Groove weld positions are for butt joints between either plate or pipe, while fillet weld positions are for T joints with plates only. As indicated in Table 7.1, some positions apply to both plate and pipe, while others apply only to plate or only to pipe.

Figures 7.15 and 7.16 show both plate and pipe positions as they would appear to the welder, and the designations for each. 1G refers to a groove weld in the flat position. Welding in the flat

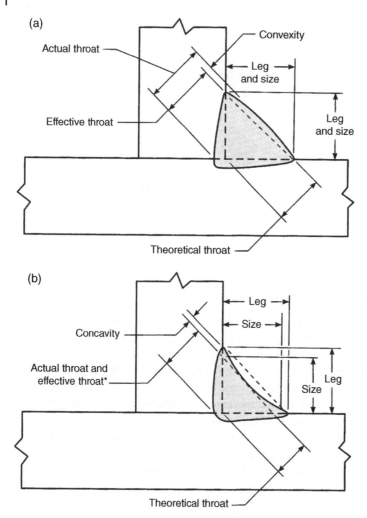

Figure 7.14 Fillet weld nomenclature. (a) Convex fillet weld and (b) concave fillet weld. (*Source:* Reproduced by permission of American Welding Society, ©*Welding Handbook*).

Table 7.1 Designation for weld position/type and their descriptions.

Position or weld type	Description
1	Flat—can apply to plate or pipe—pipe is rotated in horizontal position
2	Horizontal—can apply to plate or pipe—pipe is in vertical position
3	Vertical—applies to plate only
4	Overhead—applies to plate only
5	Applies to pipe only: 360° weld, nonrotating pipe in horizontal position
6	Applies to pipe only: 360° weld, nonrotating pipe set at a 45° angle
F	Fillet weld—applies to plate only (T joints)
G	Groove weld—applies to plate or pipe

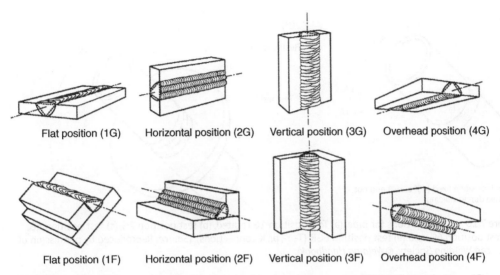

Flat position (1G)   Horizontal position (2G)   Vertical position (3G)   Overhead position (4G)

Flat position (1F)   Horizontal position (2F)   Vertical position (3F)   Overhead position (4F)

Figure 7.15 Weld positions for plate–groove welds shown on top row, fillet welds on bottom. (*Source: Welding Essentials*, Second Edition).

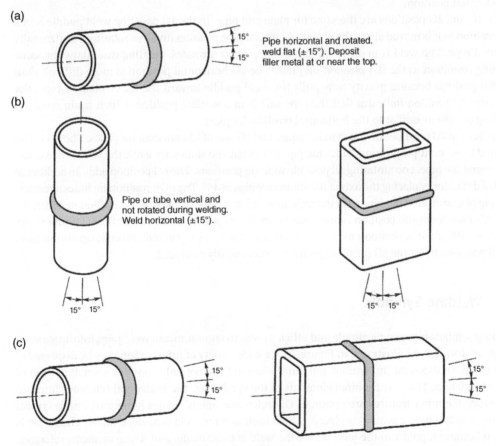

(a)

15°
15°
Pipe horizontal and rotated.
weld flat (± 15°). Deposit
filler metal at or near the top.

(b)

Pipe or tube vertical and
not rotated during welding.
Weld horizontal (±15°).

15° 15°

15° 15°

(c)

15°
15°

15°
15°

Pipe or tube horizontal fixed (±15°) and not rotated during welding.
Weld flat, vertical, overhead.

Figure 7.16 Weld positions for pipe. (a) Test position 1G rotated, (b) test position 2G, (c) test position 5G,

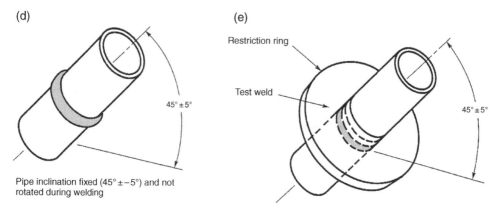

(d)

(e)

Restriction ring

Test weld

$45° ± 5°$

$45° ± 5°$

Pipe inclination fixed (45° ± −5°) and not rotated during welding

Figure 7.16 Weld positions for pipe. (a) Test position 1G rotated, (b) test position 2G, (c) test position 5G, (d) test position 6G, and (e) test position 6GR (T-, Y-, or, K-connections). (*Source:* Reproduced by permission of American Welding Society, ©*Welding Handbook*).

position whenever practical can help to improve productivity. The horizontal position is also preferred to overhead and vertical welding but is more prone to overlap and undercut defects than the flat position.

The 1G and 2G positions are the same for plate and pipe. In the 1G case, the weld puddle is in a flat position as it is moved along the groove, so the 1G pipe position involves rotating a horizontally oriented pipe. The weld is made at the top of the pipe as it rotates, creating essentially the same welding condition as the flat position on plate. The 2G horizontal position is more difficult than the flat position because gravity now pulls the weld puddle toward the lower plate or pipe. For pipe, the 2G position indicates that the pipe axis is in a vertical position, which again creates a welding condition similar to the horizontal position for plate.

The 3G and 4G positions do not exist for pipes, and 5G and 6G do not exist for plates. The 5G is also referred to as "multiple position" since the pipe axis is flat and stationary while the welder moves the arc around the pipe, encountering all possible welding positions. The 6G position adds an additional level of difficulty by placing the axis of the stationary pipe at 45°. The 6GR position includes a restricting ring placed around the pipe near the weld to simulate a typical real-world condition in which the welder's movement and position is restricted by another pipe connection. Since the 6G positions are the most difficult, it is common for welding codes to automatically qualify procedures and welders for all positions if just the 6G (or 6GR) position is successfully produced.

7.5 Welding Symbols

Welding symbols represent a simple and efficient way to communicate weld joint information on a print or drawing. As illustrated in Figure 7.17, a wide variety of information may be displayed at the various locations on the symbol, but quite often the symbol only contains a small portion of this information. The only required elements of the symbol are the horizontal reference line and the arrow; the other features are optional. The reference line is always horizontal and is critical because this is where a symbol is placed, which indicates the weld type to be made. The arrow is critical because it points to the joint where the weld is to be made, and it also creates a reference point for the weld type information that is placed on the reference line.

The symbols representing the possible weld types are shown in Figure 7.18. These symbols are known as weld symbols and represent an important part of the overall welding symbol.

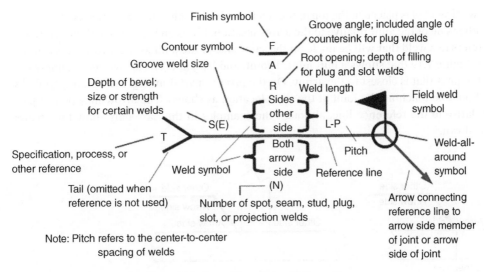

Figure 7.17 Welding symbols can include a wide variety of information.

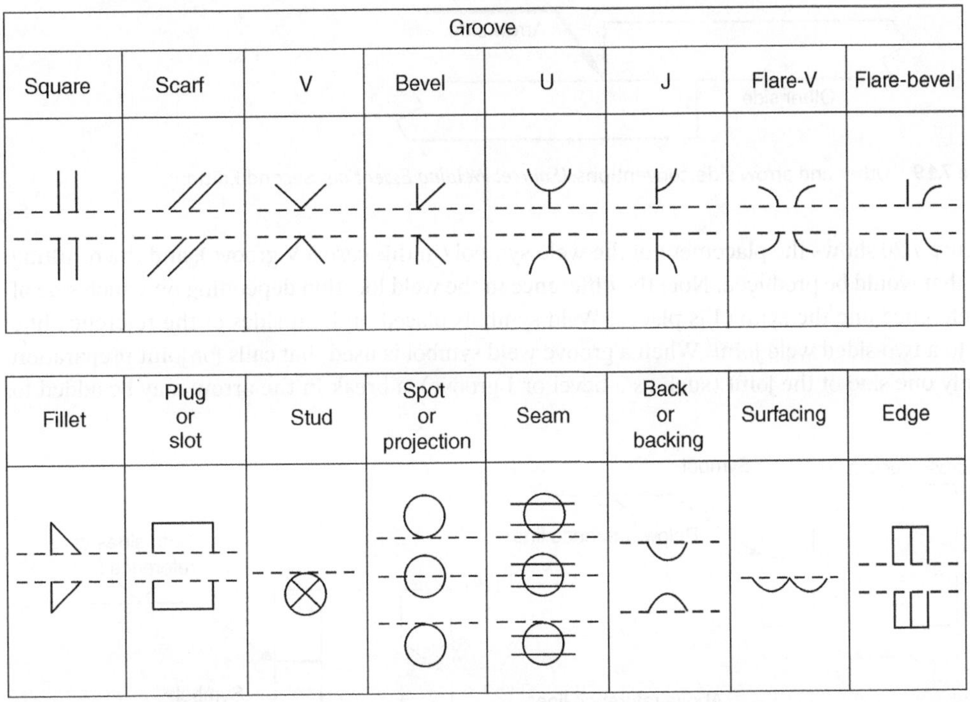

Groove							
Square	Scarf	V	Bevel	U	J	Flare-V	Flare-bevel

Fillet	Plug or slot	Stud	Spot or projection	Seam	Back or backing	Surfacing	Edge

Figure 7.18 Weld symbols and their relative position to the welding symbol reference line. (*Source: Welding Essentials*, Second Edition).

The dotted lines indicate the position of the welding symbol reference line relative to the weld symbol. To clarify, the weld symbol should not be confused with the welding symbol. The welding symbol is the entire symbol shown in Figure 7.17, whereas the weld symbol (Figure 7.18) is a specific and important component of the welding symbol that indicates the type of weld to be used.

If the weld symbol is placed at the top of the reference line, it refers to a weld that is to be made on the other side of the joint relative to the arrow location. If it is on the bottom of the reference line, it refers to a weld that will be made on the same side of the joint as the arrow. This is illustrated in Figure 7.19 that shows a butt joint, T joint, and lap joint. The arrow side is simply the side of the joint that is closest to the arrow. On the print or part drawing, the welding symbol's arrow should be touching the joint at the weld location as shown. The arrow may point up or down relative to the reference line, although the examples in this case only show the arrows pointing down.

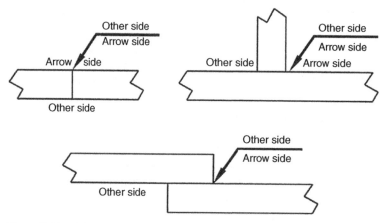

Figure 7.19 Other and arrow side conventions. (*Source: Welding Essentials*, Second Edition).

Figure 7.20 shows the placement of the weld symbol (in this case a V groove), and the resulting weld that would be produced. Note the difference in the weld location depending on which side of the reference line the symbol is placed. Weld symbols placed on both sides of the reference line refer to a two-sided weld joint. When a groove weld symbol is used that calls for joint preparation on only one side of the joint (such as a bevel or J groove), a break in the arrow may be added to

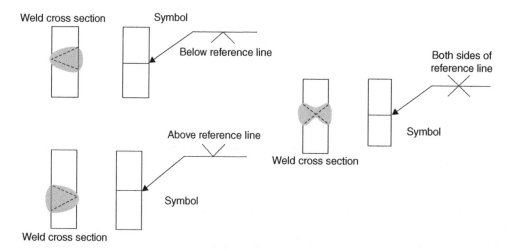

Figure 7.20 Examples of groove weld locations depending on the weld symbol position on the reference line.

orient the arrow in the direction of the side of the joint where the groove will be placed (Figure 7.21). There may be no break in the arrow line for weld types requiring joint preparation on one side only; this simply means that it either does not matter which side the joint preparation takes place, or that the groove location is obvious.

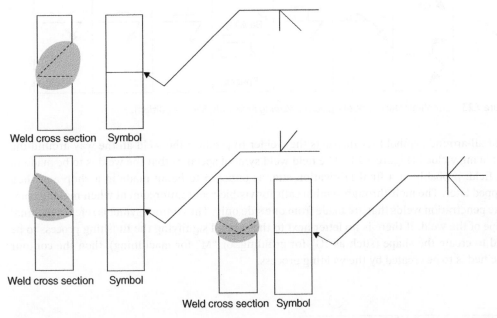

Figure 7.21 A break in the arrow line orients the arrow in the direction of the side of the joint to receive the groove preparation.

When using weld symbols that are not symmetrical such as a bevel, fillet, or J groove weld, the vertical member is always on the left side, regardless of the arrow direction (Figure 7.22). This figure also shows symbols with an optional tail, which can contain a variety of information, such as the weld process or material being welded. Supplementary symbols may be added to the arrow line (Figure 7.23). One example of the use of a supplementary symbol is the

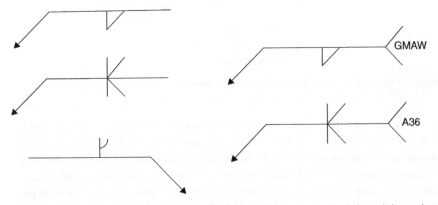

Figure 7.22 The vertical member of the weld symbol is always on the left, and the optional tail may include information such as weld process and material type.

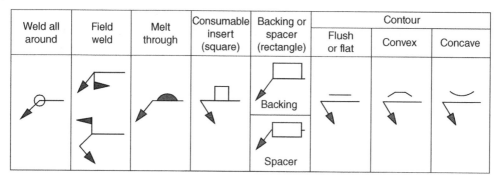

Weld all around	Field weld	Melt through	Consumable insert (square)	Backing or spacer (rectangle)	Contour		
					Flush or flat	Convex	Concave

Figure 7.23 Supplementary symbols. (*Source: Welding Essentials*, Second Edition).

weld-all-around symbol that instructs the welder to produce the weld all the way around the part being welded (Figure 7.24). The field weld symbol specifies that the weld is to be made in the field, probably at a final fabrication site, as opposed to being made in a shop and then shipped later. The melt-through symbol calls for visible root reinforcement when making complete penetration welds that are made from one side only. The contour symbols refer to the final shape of the weld. If there is no letter next to the symbol signifying the finishing process to be used to create the shape (such as "G" for grinding, or "M" for machining), then the contour specified is to be created by the welding process.

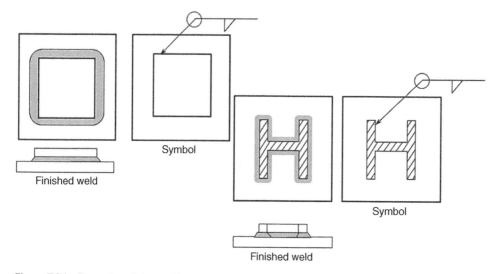

Figure 7.24 Examples of the application of the weld-all-around symbol.

Welding symbols may include information about dimensions such as root opening, bevel angle, length and pitch of intermittent welds, and weld sizes in the case of fillet welds. The locations for these various dimensions are shown in Figure 7.25. Multiple welds or multiple operations may be combined on a welding symbol as multiple reference lines or multiple symbols stacked on each other (Figure 7.26). In the case of multiple reference lines, the line closest to the arrow is performed first. In the case of multiple stacked weld symbols, the symbol closest to the reference line is performed first.

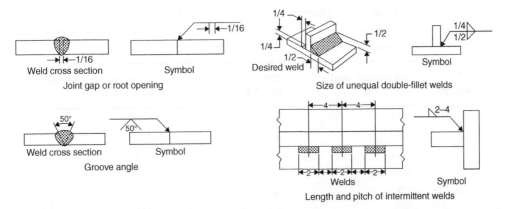

Figure 7.25 A few examples of dimensional information on welding symbols. (*Source: Welding Essentials*, Second Edition).

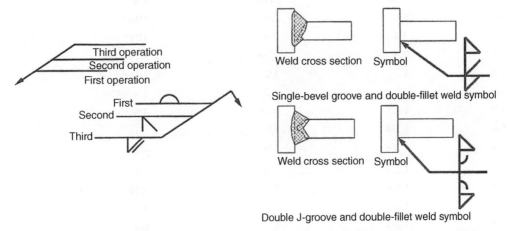

Figure 7.26 Approaches to combining welding symbols. (*Source: Welding Essentials*, Second Edition).

7.6 Weld Sizing

For an arc weld to perform well in service, it must not only be properly made but it must be properly sized. This is of particular importance with fillet welds. Since a fillet weld usually does not completely penetrate the joint, its strength is directly related to its size, and therefore, its size is very important. A fillet weld that is too small for a given loading condition may fail catastrophically. On the other hand, a weld that is too large will escalate the welding costs unnecessarily, in addition to putting excessive heat into the part or fabrication. For full penetration groove welds, there is usually no reason to be concerned about the size, but for partial penetration groove welds size may be an important consideration. Safety factors play a large role in weld sizing. For example, if a failed weld has the potential to cause injury (or worse), safety factors will be high, resulting in the need for larger weld sizes. In summary, the process of determining proper weld sizes is often a critical function of the Design Engineer, but something the Welding Engineer should be very familiar with. And as mentioned, weld sizing is primarily an issue with fillet welds.

There are some general "rule of thumb" approaches that can provide guidance regarding the selection of fillet weld sizes. For example, Table 7.2 shows typical minimum fillet weld sizes as a function of plate thickness and loading condition. In this case, strength design refers to situations

Table 7.2 Minimum fillet weld sizes as a function of plate thickness and design criteria.

Plate thickness (*t*)	Strength design	Rigidity design	
	Full strength weld ($\omega = 3/4\ t$)	50% of full strength weld ($\omega = 3/8\ t$)	33% of full strength weld ($\omega = 1/4\ t$)
1/4	3/16	3/16*	3/16*
5/16	1/4	3/16*	3/16*
3/8	5/16	3/16*	3/16*
7/16	3/8	3/16	3/16*
1/2	3/8	3/16	3/16*
9/16	7/16	1/4	1/4*
5/8	1/2	1/4	1/4*
3/4	9/16	5/16	1/4*
7/8	5/8	3/8	5/16*
1	3/4	3/8	5/16*
1 1/8	7/8	7/16	5/16
1 1/4	1	1/2	5/16
1 3/8	1	1/2	3/8
1 1/2	1 1/8	9/16	3/8
1 5/8	1 1/4	5/8	7/16
1 3/4	1 3/8	3/4	7/16
2	1 1/2	3/4	1/2
2 1/8	1 5/8	7/8	9/16
2 1/4	1 3/4	7/8	9/16
2 3/8	1 3/4	1	5/8
2 1/2	1 7/8	1	5/8
2 5/8	2	1	3/4
2 3/4	2	1	3/4
3	2 1/4	1 1/8	3/4

Source: Adapted from *Design of Weldments*, The James F. Lincoln Arc Welding Foundation.
*Values have been adjusted to comply with AWS-recommended minimums.

where the weld is expected to carry the load being placed on the welded fabrication. Rigidity design applies when the weld is not required to directly carry the load, such as when a weld is simply holding a stiffener in place.

The "rule of thumb" approaches are only provided as a guide, and quite often weld sizes must be calculated based on specific loading conditions and required safety factors. A common and relatively simple approach to determining fillet weld sizes uses a concept known as "treating the weld as a line" which involves calculations based on weld length and not area. Instead of calculating a weld stress (lbs/in.^{2}) based on the cross-sectional area of the weld, a unit force per unit length (lbs/in.) is determined based on the type and magnitude of loading, and the length and geometry of the weld.

Once this is determined, equations based on a unit force (Table 7.3) can be used to calculate the stress on the weld, and this number can then be compared to an allowable stress. Allowable

Table 7.3 Equations for the calculation of force per unit length of weld.

Type of loading		Standard design formula stress lbs/in.^{2}	Treating the weld as a line force lbs/in.
	Primary welds Transmit entire load at this point Tension or compression	$\sigma = \dfrac{P}{A}$	$f = \dfrac{P}{A_w}$
	Vertical shear	$\sigma = \dfrac{V}{A}$	$f = \dfrac{V}{A_w}$
	Bending	$\sigma = \dfrac{M}{S}$	$f = \dfrac{M}{S_w}$
	Twisting	$\tau = \dfrac{TC}{J}$	$f = \dfrac{TC}{J_w}$
	Secondary welds Hold section together—low stress Horizontal shear	$\tau = \dfrac{V_{ay}}{I_t}$	$f = \dfrac{V_{ay}}{I_n}$
	Torsional horizontal shear	$\tau = \dfrac{T}{2At}$	$f = \dfrac{T}{2A}$

Source: Adapted from *Design of Weldments*, The James F. Lincoln Arc Welding Foundation.

stresses vary based on the weld type, loading conditions, and the required safety factor, but a common and conservative allowable stress applied to fillet welds is equal to 0.3 times the ultimate tensile strength of the filler metal. Finally, the unit force and allowable stress can be used to determine a minimum weld size.

In summary, the steps for determining a minimum weld size by treating the weld as a line are as follows:

1) Determine loads and loading conditions on the weld.
2) Measure the total weld length.
3) Using the weld geometry, select appropriate section properties (Table 7.3) to determine the unit force on the weld.
4) Use the standard formula for stress ($\sigma = F/A$) and compare to allowable stress to determine leg size of weld (where F becomes a unit force F_r which is equal to the load divided by the total weld "area" or weld length, and $A = 0.707 \times$ leg size (t_w)).
5) Solve for leg size (t_w), and then round up to the nearest 1/16″ increment.

The reason for the multiplying factor of 0.707 in the denominator is to convert leg size to the effective throat dimension, which is the smallest "ligament" of weld metal supporting the load. However, the leg size is what is important since it can easily be measured by the Welding Inspector using an appropriate gauge (whereas the throat dimension can only be measured metallographically). The final step involves rounding up the weld leg size (t_w) to the nearest 1/16 in. increment. The reason for this is because ultimately the weld size shown on the print or drawing (which informs the welder the weld size that is required) will have to be in 1/16 in. increments since the weld size gauges used by inspectors are designed to measure in those increments.

An example of a fillet weld size calculation for a simple loading condition involving a plate welded to another plate or structure is shown on Figure 7.27. It should be pointed out that this is just one simple example, and that there are other design approaches to determining weld sizes. Also, these calculations become much more complex when other loading types such as bending, twisting, and/or other conditions such as cyclic loading are involved.

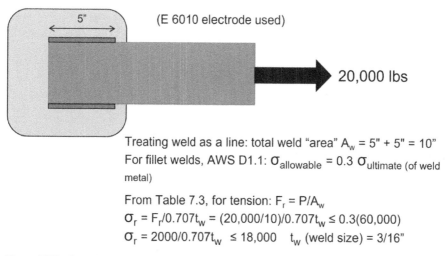

Treating weld as a line: total weld "area" $A_w = 5" + 5" = 10"$
For fillet welds, AWS D1.1: $\sigma_{allowable} = 0.3\ \sigma_{ultimate\ (of\ weld\ metal)}$

From Table 7.3, for tension: $F_r = P/A_w$
$\sigma_r = F_r/0.707t_w = (20{,}000/10)/0.707t_w \leq 0.3(60{,}000)$
$\sigma_r = 2000/0.707t_w \leq 18{,}000 \quad t_w$ (weld size) $= 3/16"$

Figure 7.27 Sample calculation of fillet weld size treating the weld as a line.

7.7 Computational Modeling of Welds

As computer technology continues to advance, the subject of computational modeling of welds continues to grow in importance. Computational modeling refers to the use of powerful computer software programs that can simulate a wide variety of welding-related phenomena. Heat flow, effect of process variables and joint designs, and residual stress and distortion are all examples of welding topics that can be simulated through process modeling software. In addition to processes, microstructures, phases, and properties can be simulated through materials modeling software. Computer simulations can provide for the ability to analyze complex welding phenomena much more quickly and efficiently as compared to more traditional experimental methods.

For example, Figures 7.28 and 7.29(a) show images of a Resistance Spot Welding simulation next to an actual spot weld made with the same parameters used for the model. In both figures, the different shades in the simulation represent different temperatures. In Figure 7.28, the dotted white line depicts what the model predicted, while the solid dark line shows where the actual weld nugget formed. In Figure 7.29(a), the light gray shade represents the molten nugget, while the solid darker line shows the predicted edge of the nugget fusion line. Figure 7.29(b) and (c) show how the predicted nugget diameter and thickness compare with the corresponding experimental values as

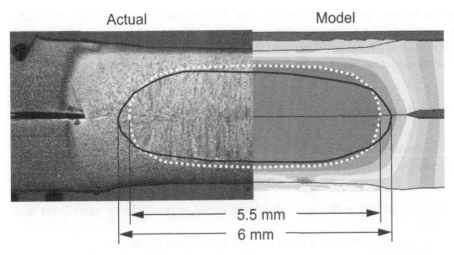

Figure 7.28 Simulated resistance spot weld next to an actual weld. (*Source:* Dr. Max Biegler, Fraunhofer Institute).

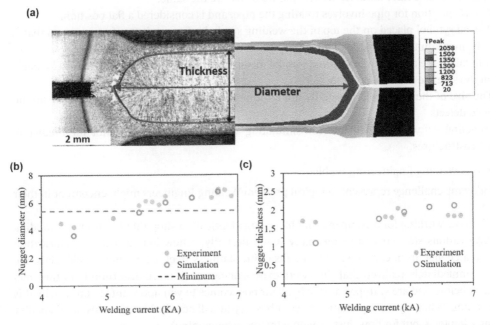

Figure 7.29 Simulated resistance spot weld next to an actual weld, and comparison of nugget diameter and thickness as a function of welding current. (*Source:* Dr. Ying Lu et al., The Ohio State University).

a function of welding current. Both figures reveal some slight variation between the model and the actual weld, but the differences are negligible. This ability to simulate heat patterns during Resistance Spot Welding provides for much more rapid development and optimization of welding parameters for various sheet thicknesses and material types. Modeling software for other welding processes exists as well.

There are many commercially available software packages for modeling both welding processes and welding metallurgy and properties. Software packages that are ready to use for welding process simulation include Sorpas, Sysweld, and Simufact, and Thermo-Calc is a materials modeling package that is also ready to use for welding metallurgy simulation. Other general-purpose finite element software packages such as Abaqus and Ansys provide for the ability to develop customized programs for welding process simulation, but these packages require advanced programming knowledge.

7.8 Test Your Knowledge

I) Fundamental Concepts—True/False

The following true/false questions pertain to some of the most important fundamental concepts in this chapter:

1) Because of its low melting temperature, aluminum can be arc welded with much lower heat input than steel.
2) Yield stress is the maximum stress a metal can withstand before it fractures.
3) Metals with high electrical conductivity are the easiest to resistance weld.
4) A big benefit of fillet welds is there is no special joint preparation required.
5) With a concave fillet weld, leg length and weld size are the same.
6) The 1G position for pipe involves rotating the pipe, and is considered a flat position.
7) A weld symbol placed on the top of the welding symbol reference line refers to a weld that is to be made on the other side of the joint the arrow is pointing to.
8) A break in the arrow of the welding symbol means that the groove preparation must be made on the side the arrow is point to.
9) The size of the fillet weld is usually not an important consideration as long as there are no weld defects.
10) Modeling software is available for simulating welding processes, but not for predicting microstructures.

II) Solve a Welding Engineering Problem

The following challenge represents a typical problem Welding Engineers might encounter in their career:

You're now working for a company that has a large fabrication shop that has been arc welding (GMAW) various steel structures for many years. Recently, a new job on a certain fabrication involved a change from carbon steel to an austenitic stainless steel. Upon initial weld trials, the welders immediately realized that they were experiencing much more distortion than they had ever seen before and are stumped as to why. Your boss comes to you and wants to know why this is happening. What is your explanation to your boss? (you will be asked to solve this problem after reading Chapter 8, but for now, just explain why this is happening).

Recommended Reading for Further Information

AWS Welding Handbook, Ninth Edition, Volume 1—"Welding Science and Technology," American Welding Society, 2001.
Design of Weldments, The James F. Lincoln Arc Welding Foundation, 1972.

8

Heat Flow, Residual Stress, and Distortion

8.1 Heat Flow

Nonuniform heat flow during welding is the primary cause of residual stress, which in turn produces distortion. Avoiding or minimizing residual stress and/or distortion in weldments often contributes to significant cost in producing the weldment. Residual stresses can be minimized through heat treatments and other means to improve dimensional stability and reduce susceptibility to problems such as hydrogen cracking and fatigue cracking. Controlling distortion may require expensive tooling and fixturing, and possibly postweld machining. Prior to discussing the fundamental mechanisms which create residual stress and the methods for mitigation, it is first important to review some basic heat flow principles.

There are three types of heat transfer: convection, radiation, and conduction (Figure 8.1). Convection refers to the transfer of thermal energy through mass movement, and is usually associated with movement of fluids, such as in the heating of a home. Radiation is the transfer of thermal energy through the emission and absorption of electromagnetic radiation, such as the heat one feels from sunshine. Conduction is the transfer of thermal energy within a single body, or from one body in contact with another body. An example is the extreme cold one would experience when placing their hand in a bucket of ice water. Conduction is the most important form of heat transfer during welding and will be discussed in more detail.

Through somewhat rigorous mathematical derivations that are covered in more advanced Welding Engineering courses, the law of heat conduction known as Fourier's Law can be used to derive the one-dimensional (1-D) heat flow equation (Figure 8.2). Fourier's Law dictates that heat flows from hot to cold regions (hence the minus sign), and the driving force is a function of the temperature gradient (slope). Steeper gradients and higher thermal conductivity of the material both result in greater heat flow per unit area. The similarities of the 1-D heat flow equation to Fick's Second Law of Diffusion (which predicts how diffusion causes the concentration of a substance to change with time) are no coincidence—the transfer of heat energy is a form of diffusion. The 1-D heat equation reveals the importance of the three physical properties of metals to the rate of heat flow: thermal conductivity (λ), density (ρ), and heat capacity (C_p). Higher thermal conductivities result in greater heat flow, while greater densities and heat capacities result in larger storage of heat energy. As shown in the equation, thermal conductivity divided by the product of density multiplied by heat capacity is known as thermal diffusivity (k).

Welding Engineering: An Introduction, Second Edition. David H. Phillips.
© 2023 John Wiley & Sons, Inc. Published 2023 by John Wiley & Sons, Inc.
Companion Website: www.wiley.com/go/Phillips/WeldingEngineeringIntroduction

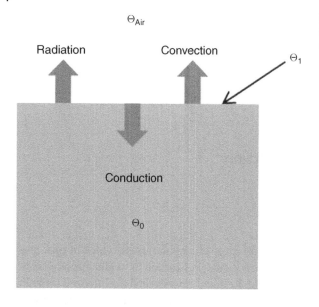

Θ_{Air}

Radiation   Convection Θ_1

Conduction

Θ_0

Figure 8.1 The three types of heat transfer: convection, radiation, and conduction.

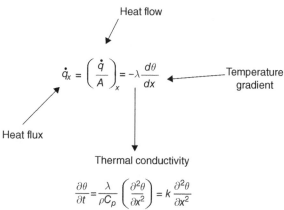

Heat flow

$$\dot{q}_x = \left(\frac{\dot{q}}{A}\right)_x = -\lambda\frac{d\theta}{dx}$$

Temperature gradient

Heat flux

Thermal conductivity

$$\frac{\partial\theta}{\partial t} = \frac{\lambda}{\rho C_p}\left(\frac{\partial^2\theta}{\partial x^2}\right) = k\frac{\partial^2\theta}{\partial x^2}$$

Where k = thermal diffusivity

Figure 8.2 Fourier's law of heat conduction (top), and the 1-D heat flow equation.

In practice, when considering heat flow in a welding application, thermal conductivity is the property that is commonly referred to as having the biggest impact on welding heat flow and cooling rates. This is because thermal conductivity tends to be the dominant factor in the equation for thermal diffusivity when evaluating commonly welded structural metals. For example, the thermal conductivity of aluminum is about three times greater than that of iron. When incorporating density and heat capacity in the thermal diffusivity equation, aluminum's thermal diffusivity is more than four times greater than iron. In both cases the differences are significant, and it is well known that aluminum extracts heat from the weld region at a much greater rate than iron (or steel). So, it is common to conclude that this difference in cooling rates between aluminum and steel is due to the difference in thermal conductivity, although technically, it is really a difference in thermal diffusivity.

While the 1-D heat flow equation clearly establishes the importance of the thermal conductivity and thermal diffusivity of the material being welded, heat flow in weldments can be categorized as either two-dimensional (2-D) for relatively thin sections, or three-dimensional (3-D) for relatively thick sections (Figure 8.3). Mathematical solutions for heat flow in weldments are complex and beyond the scope of this book; however, if an oval weld shape and point heat source are assumed, relatively simple equations can be developed and used to estimate weld heat flow (or cooling rate) in 2-D and 3-D situations (Equation 8.1: 2-D heat flow equation at top and 3-D equation at bottom— source: *AWS Welding Handbook*, Vol. 1, Ninth Edition).

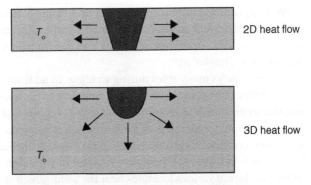

Figure 8.3   2-D versus 3-D heat flow in weldments.

2-D :

$$R_C = -2\pi k \rho C \left(h / H_{net} \right)^2 \left(T_C - T_O \right)^3$$
$$H_{net} = fVI / S$$

where

R_C = cooling rate at weld centerline

k = thermal conductivity of the base metal

ρ = density of the base metal

C = specific heat of the base metal

h = base metal thickness

H_{net} = net heat input per unit length

T_C = temperature at which the cooling rate is calculated

T_O = initial temperature of the base metal

f = efficiency

V = volts

I = amps

S = weld travel speed

(8.1)

3-D :

$$R_C = -2\pi k \left(T_C - T_O \right)^2 / H_{net}$$

As discussed, heat flow during welding depends on many factors, but the thickness of the plate (or part) plays a big role. For relatively thin parts, heat flow is 2-D—it flows in the two directions parallel to the plane of the plate. A plate is considered thin when the difference in temperature between the top and the bottom of the plate near the weld is negligible. The 3-D equation is used for relatively thick plates. A plate is considered thick when there is a large difference in temperature between the top and bottom of the plate in regions close to the weld. Another rule of thumb for 3-D heat flow is that 3-D conditions apply to weldments that require greater than four passes. Both equations can be used to estimate the weld cooling rate at the center of the weld for different metals and weld heat inputs. Calculations using these equations reveal that 3-D heat flow results in faster cooling rates, which increases the chance of forming brittle martensite when welding certain steels. These equations also show the importance of the starting temperature (T_0) of the material being welded, explaining a primary reason why preheating is often needed prior to welding. By increasing T_0 (preheating), weld cooling rates can be significantly reduced, and brittle martensite formation can be controlled. Brittle martensite is a major contributor to hydrogen cracking, a topic that is addressed in Chapter 10.

Heat flow plays many roles during welding. In addition to the issue of contributing to the creation of brittle martensite in steel, rapid heat flow in metals that exhibit high thermal conductivities (or diffusivities) such as aluminum and copper require additional heat input to create a weld. Poor heat flow in materials such as austenitic stainless steels, when combined with a high thermal expansion coefficient, can contribute to excessive distortion.

As shown in Figure 8.4, heat flow in a weldment determines both peak temperatures and time at temperature at various locations near the weld, which in turn affects the size of the heat-affected zone. Higher heat input welding processes will produce higher peak temperatures at given distances from the weld, slower cooling rates, and larger heat-affected zones (Figure 8.5). The weld heat-affected zone will be discussed in more detail in subsequent chapters.

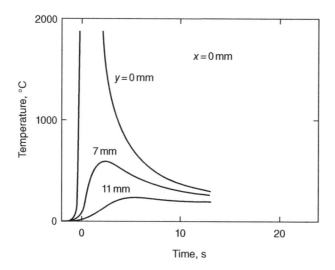

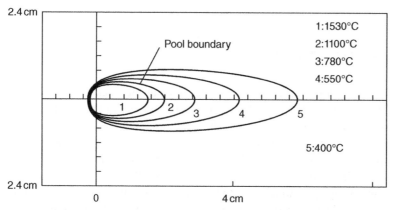

Figure 8.4 Peak temperatures, time at temperature, and cooling rates at various distances from a typical weld fusion line. (*Source: Welding Metallurgy*, Second Edition Figure 2.18, page 52. Reproduced with permission from John Wiley & Sons).

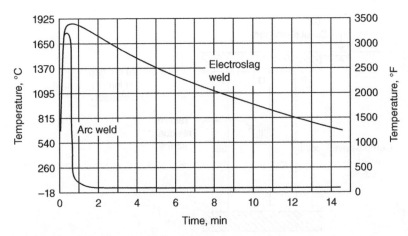

Figure 8.5 Cooling rate comparison between a typical arc weld and an extremely high heat input process such as an electroslag weld. (*Source: Welding Metallurgy*, Second Edition Figure 2.24, page 56. Reproduced with permission from John Wiley & Sons).

8.2 Fundamentals and Principles of Residual Stress and Distortion

Residual stress refers to the stress that exists in a weldment after all external loads are removed, and distortion is the dimensional change or warpage that occurs in a weldment due to residual stress. Residual stresses and distortion are the result of the nonuniform heating that is typical of most welding processes such as arc welding. To understand the fundamental mechanisms that contribute to the formation of residual stresses and distortion, it is helpful to first study what is known as a "three-bar analogy."

As depicted in Figure 8.6, the three-bar analogy considers a scenario where two outer bars of equal length are rigidly connected with a split bar between them. If the gap in the split bar is forcibly closed, a tensile stress will be created in the split bar and compressive stresses will be created in the outer bars. Consider another similar scenario in which all three bars are solid and of equal length. What would happen to the stress level of the center bar as it is heated and then cooled? Referring to the plot of Figure 8.7, upon heating, the center bar attempts to expand and lengthen but is rigidly constrained due to the connection to the other bars. This results in the initial development of compressive stresses (A-B). But as the bar continues to heat, its yield strength continues to drop causing the compressive stresses to drop accordingly as the bar yields at lower and lower stress levels (B-C). The stress level in the bar eventually becomes negligible since the yield strength becomes very low at elevated temperatures. This remains the case as the bar is heated higher and then initially cooled (C-D-E).

But at some point, upon further cooling, the yield strength of the bar begins to rapidly increase while the bar attempts to shrink. Since the bar at this point is essentially free of compressive stresses due to the previous yielding, tensile stresses can begin to build as the bar shrinks while under constraint. As cooling continues, the bar attempts to shrink further while yield strength continues to increase resulting in further increases in tensile stresses (E-X-G). Upon final cooling, the tensile stresses remaining in the bar will be near or equivalent to the room temperature yield strength of that material. This stress is what is known as residual stress.

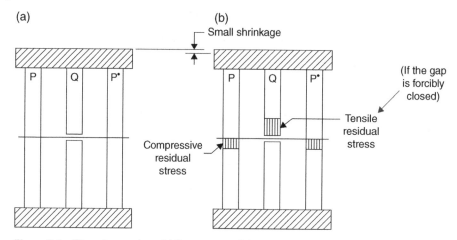

Figure 8.6 Three-bar analogy. (a) Free state and (b) stressed state. (*Source:* Reproduced by permission of American Welding Society, ©*Welding Handbook*).

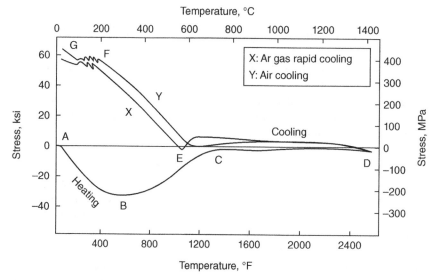

Figure 8.7 Stress level of middle bar as it is heated and then cooled. (*Source:* Reproduced by permission of American Welding Society, ©*Welding Handbook*).

When comparing the three-bar analogy to welding, one can think of the center bar as the heated weld region, while the outer bars are analogous to the cooler base metal surrounding the weld. Now consider a bead-on-plate arc weld (Figure 8.8). As the weld is formed, the heated metal surrounding the weld region attempts to expand. Some expansion occurs causing the plates to distort slightly in one direction, but the expansion is mostly restricted by the surrounding cool metal. Since the surrounding cool metal is much stronger than the hot metal (and of course the liquid weld metal has no strength), yielding of the hot metal occurs allowing the region to return to a relatively low stress state. Upon cooling, the liquid metal solidifies and shrinks (solid metal has less volume than liquid metal), as does the surrounding heated metal. As the shrinkage continues, the heated metal continues to cool and regain yield strength and begins to pull like a spring on the

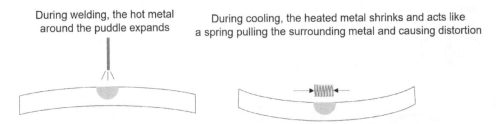

During welding, the hot metal around the puddle expands

During cooling, the heated metal shrinks and acts like a spring pulling the surrounding metal and causing distortion

Figure 8.8 The development of residual stress and distortion when welding.

surrounding cooler metal causing it to strain or distort in the other direction. The permanent strain that remains is known as distortion, and the permanent stress is known as residual stress. As with the three-bar analogy, residual stresses that remain can be as high as the room temperature yield strength of the material.

Residual stress patterns in weldments can be quite complex, but always consist of a balance of tensile and compressive stresses to create a net force of zero. Tensile stresses typically exist at and near the weld, while the compressive stresses are found in the surrounding base metal. A typical and very simple residual stress pattern is shown in Figure 8.9. Residual stresses and their patterns vary greatly and are primarily a function of the amount of heat input, the joint design, and the degree of restraint.

To further understand how distortion occurs, it is helpful to review the plot of Figure 8.10, which is a plot of the amount of deflection of the welded plate versus time. The initial deflection (A-B)

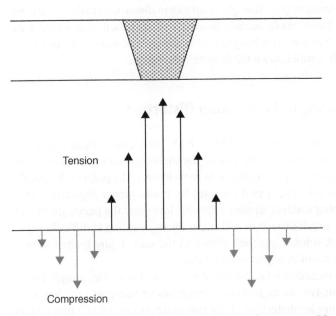

Tension

Compression

Figure 8.9 Typical residual stress pattern in a simple groove weld.

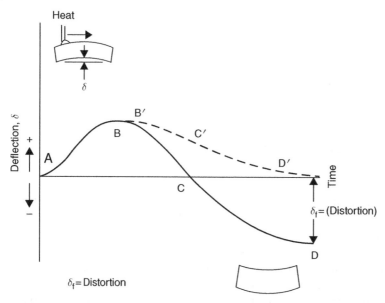

Figure 8.10 Cause of distortion. (*Source:* Reproduced by permission of American Welding Society, ©*Welding Handbook*).

occurs due to the expanding hot metal (left side of Figure 8.8) pushing on the surrounding plate. If all the movement was elastic, upon cooling, the plate would be expected to return to its original state and follow path B′-C′-D′. However, because significant yielding is also occurring as the heated metal is attempting to expand against the much stronger cooler metal, upon further cooling, the shrinking weld region continues to gain strength resulting in distortion (path B-C-D). As mentioned previously, metals with higher coefficients of thermal expansion will expand more as they are heated and shrink more as they cool, resulting in higher amounts of distortion. An example is austenitic stainless steel which is well known for distortion problems.

8.3 Approaches to Minimizing or Eliminating Distortion

A few of the many types of distortion are shown in Figure 8.11. Fortunately, there are many approaches for minimizing or eliminating distortion. One of the most obvious ways is known as presetting (Figure 8.12) in which the joint angle is offset prior to welding. If the proper offset angle is chosen, the subsequent distortion will simply pull the joint back into proper alignment. This approach has the advantage of keeping residual stresses relatively low since the plates are free to move. But anticipating exactly how much distortion will occur during welding is difficult, especially if the fabrication is complex. A similar approach shown in the same figure known as pre-straining is easier to control but will result in higher residual stresses.

Other forms of distortion control include a wide variety of methods of restraint. Simple forms of clamping are most common, but two more unique approaches to restraint are shown in Figure 8.13. In the top approach, angular distortion of the two plates is restricted while distortion in other directions, such as transverse shrinkage, is allowed to occur. This approach helps keep residual stresses to a minimum when there is no concern for distortion that occurs in any direction other than the one that is being restrained. Another unique approach shown when

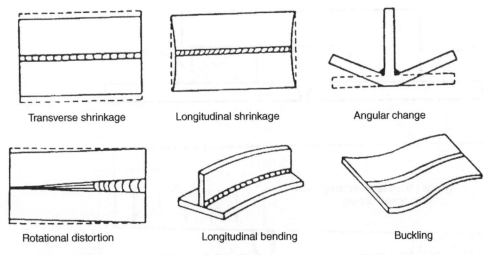

Transverse shrinkage   Longitudinal shrinkage   Angular change

Rotational distortion   Longitudinal bending   Buckling

Figure 8.11 Typical examples of forms of distortion.

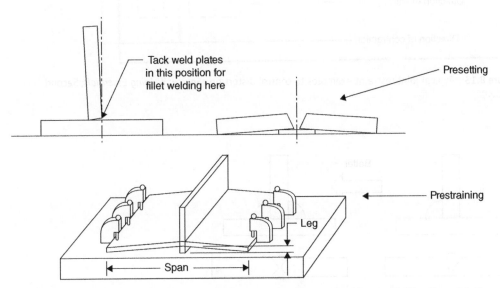

Tack weld plates in this position for fillet welding here

Presetting

Prestraining

Leg

Span

Figure 8.12 Methods of distortion control, presetting and prestraining. (*Source: Welding Essentials*, Second Edition).

welding plates in a butt joint configuration is to place a wedge along the joint to keep the joint from closing during welding. However, it is much more common to simply place tack welds on each end of the joint prior to creating the weld.

Joint design and/or weld sequencing is often the easiest way to control distortion. Welds made on both sides of the joint will distort less than single-sided welds (Figure 8.14). Approaches to weld sequencing in the case of fillet welds include chain or intermittent staggered welds (Figure 8.15) in which segments of welds are made alternatively from one side to the other. Intermittent welding also reduces overall heat input, which minimizes distortion even further. These intermittent fillet weld approaches are only an option if sufficient weld strength can be achieved without a continuous weld. The back-step welding approach for groove welds (Figure 8.16) to control distortion in a

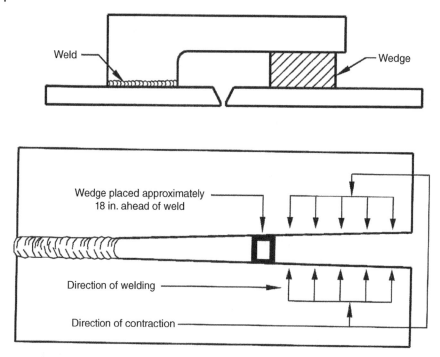

Figure 8.13 Typical joint restraint examples to control distortion. (*Source: Welding Essentials*, Second Edition).

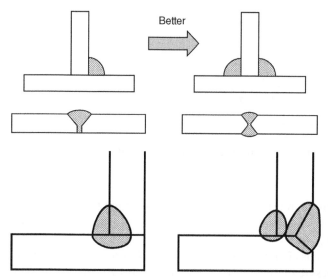

Figure 8.14 Welding on both sides of the joint will significantly reduce distortion. (*Source: Welding Essentials*, Second Edition).

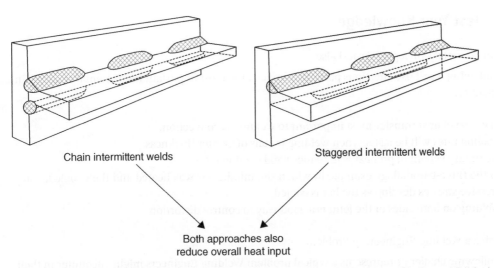

Chain intermittent welds

Staggered intermittent welds

Both approaches also
reduce overall heat input

Figure 8.15 Intermittent fillet welds to control distortion. (*Source: Welding Essentials*, Second Edition).

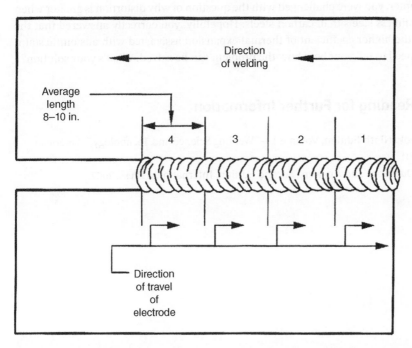

Figure 8.16 Distortion control of a groove weld through back-step welding. (*Source: Welding Essentials*, Second Edition).

long continuous bead involves producing segments of welds in individual steps to produce a continuous weld. In this case, weld segment #1 is initiated with electrode travel from left to right, followed by weld segment #2, and so on. Of course, this approach will increase the time it takes to complete a weld.

8.4 Test Your Knowledge

I) Fundamental Concepts—True/False

The following true/false questions pertain to some of the most important fundamental concepts in this chapter:

1) The type of heat transfer most important to welding is convection.
2) Cooling rates will be faster when welding metals of greater thickness.
3) A primary benefit of preheat is to reduce weld cooling rates.
4) In the three-bar analogy example in which the middle bar was heated and then cooled, compressive stresses develop as the bar is cooled.
5) Welding on both sides of the joint is a good way to control distortion.

II) Solve a Welding Engineering Problem

The following challenge represents a typical problem Welding Engineers might encounter in their career:

    In the previous chapter, you were challenged with the question of why distortion is greater when welding austenitic stainless steels than carbon steels. Hopefully, you correctly answered that the reason is because of the higher coefficient of thermal expansion associated with austenitic stainless steels. Now you need to tell your boss how the problem can be solved. What's your solution?

Recommended Reading for Further Information

AWS Welding Handbook, Ninth Edition, Volume 1—"Welding Science and Technology," American Welding Society, 2001.

Welding Essentials—Questions and Answers, Second Edition, International Press, 2007.

9

Welding Metallurgy

9.1 Introduction to Welding Metallurgy

Welding metallurgy refers to the complex microcosm of metallurgical processes that can occur in and around a weld during the rapid heating and cooling cycles associated with most welding processes. Because welding always occurs under nonequilibrium conditions in which diffusion is limited, many of the standard metallurgical principles that exist either cannot be applied or can only be used as an approximation of metallurgical behavior. As such, a body of knowledge known as welding metallurgy has evolved to describe the microstructure evolution and associated properties of welds. Welding metallurgy is complex because a single weld can be comprised of a wide variety of metallurgical processes, depending on the metal being welded, how it was processed, and the type of welding process. The metallurgical processes that occur are important to understand since they directly affect the microstructure that is created, and therefore, influence the subsequent mechanical properties of the joint. In most cases, the microstructural changes that occur due to welding result in a degradation of base metal properties.

Melting and solidification are two obvious and important metallurgical processes associated with the region known as the fusion zone produced when welding with fusion welding processes. Aspects of weld solidification include the nucleation and growth of dendrites (Figure 9.1), segregation, and diffusion processes resulting in localized compositional variations. Such compositional variations may create welding problems, as well as influence performance in service.

The weld heat-affected zone (HAZ) can be loosely defined as the entire region outside the fusion zone in which the base metal properties of the metal being welded have been affected in some way. Many solid-state metallurgical processes may occur in the HAZ such as phase transformations, diffusion, precipitation reactions, recrystallization, and grain growth. Liquation reactions, in which liquid films may form just outside of the fusion zone, can result in cracking problems. The extent of these reactions may significantly alter the microstructure and properties of the weldment relative to the base metal. Many of these reactions, or combinations of reactions, can result in embrittlement of welds.

As Figure 9.2 indicates, to anticipate and understand the mechanical properties of a weldment and how it will perform in service, it is important to first understand the microstructure that is created during welding. To understand the weld microstructure, and in particular the weld HAZ, it is important to first have a full understanding of the base metal microstructure and how it was processed, as well as the type of welding process being used. For example, steel can be strengthened by what is known as transformation hardening in which a hard and strong microstructure

Welding Engineering: An Introduction, Second Edition. David H. Phillips.
© 2023 John Wiley & Sons, Inc. Published 2023 by John Wiley & Sons, Inc.
Companion Website: www.wiley.com/go/Phillips/WeldingEngineeringIntroduction

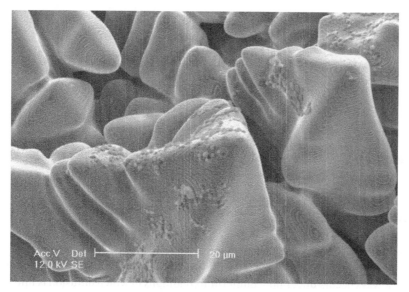

Figure 9.1 Dendritic solidification. (*Source:* Dr. John Lippold, The Ohio State University).

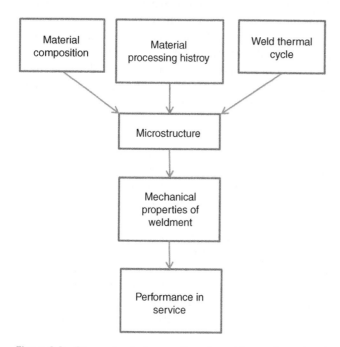

Figure 9.2 The mechanical properties of a weldment depend on its microstructure, which depends on the material being welded and its processing history as well as the welding process.

known as martensite is intentionally created, and then its properties are customized through a tempering heat treatment. When such a steel is welded, it is very likely that untempered martensite will be formed, which can lead to cracking and/or loss of ductility and toughness. High-energy density welding processes such as Laser Welding may be more likely to form martensite than a process such as Submerged Arc Welding, which is known for high heat input and slow cooling rates. An aluminum alloy that was previously strengthened via precipitation hardening may be subject to severe softening in the HAZ through an overaging reaction. In this case, a high-energy density welding process such as Laser Welding might be beneficial since it is a low heat input process.

Terminologies used to describe the regions of a fusion weld include the fusion zone, which consists of the composite zone (CZ) and unmixed zone (UMZ), and heat-affected zone, which consists of the partially melted and true heat-affected zone or HAZ (Figure 9.3). Little has changed since 1976 regarding the terminologies for describing these, although considerable research has been conducted on a variety of alloy systems to verify that these regions are valid. Additional refinements have been made to this original terminology. For example, the true HAZ in steels has been divided into various subregions, such as the coarse-grained HAZ, the fine-grained HAZ, and the intercritical (in which peak temperatures reach the alpha ferrite + austenite phase field) region. Transition zones may be created in dissimilar metal weldments such as in welds between stainless steels and low alloy steels where martensite may form in the transition region even though it does not occur elsewhere in the weld.

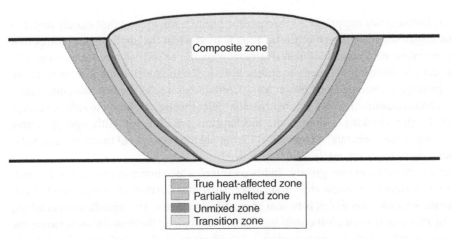

Figure 9.3 Terminology describing the regions of a fusion weld. (*Source:* Dr. John Lippold, The Ohio State University).

9.2 The Fusion Zone

The fusion zone represents the region of a fusion weld where there is complete melting and resolidification during the welding process. It is usually metallographically distinct from the surrounding HAZ and base metal. The microstructure in the fusion zone is a function of the alloy composition and solidification conditions. For example, rapid cooling rates will result in more rapid solidification and a finer fusion zone microstructure. Although it is customary to refer to the entire region of melting as the fusion zone, in welds where the filler metal is of a different composition than the base metal, three regions can hypothetically exist. The largest of these is the composite zone, consisting of filler metal diluted with melted base metal. Adjacent to the fusion boundary, two additional regions may exist. The unmixed zone consists of melted and resolidified base metal where negligible mixing with filler metal has occurred.

Between the unmixed and composite zone, a transition zone must exist where a composition gradient from the base metal to the composite zone is present. As mentioned previously, a transition zone may be especially important in a dissimilar metal weld. Also, fusion zones can be divided into three general categories—autogenous, homogeneous, and heterogeneous. These classifications are based on whether filler metal is added, and if so, does it match the base metal composition.

Autogenous (no filler metal added) welding is most common when welding thin sections. It can often be conducted at high speeds and with a minimum amount of joint preparation. Butt joints may be used, but edge welds are often ideal for autogenous welding. Joint fit-up should be very tight since there is no filler metal to account for gaps in the weld joint. Although there are many welding processes such as Electron Beam and Resistance Welding that do not use a filler metal, the term "autogenous" is reserved for welds created with the Gas Tungsten and Plasma Arc Welding processes. The fusion zone is essentially of the same composition as the base metal, except for possible compositional changes due to metal evaporation or absorption of gasses from the surrounding atmosphere. Not all materials can be joined autogenously because of weldability issues such as solidification cracking, a topic discussed in Chapters 11 and 12.

Homogenous fusion zones are produced through the use of a filler metal that closely matches the base metal composition. This type of fusion zone is used when the application requires that filler and base metal properties, such as heat treatment response or corrosion resistance, must be closely matched. Quite often it is necessary to choose a dissimilar filler metal composition from the base metal to produce a heterogeneous fusion zone. Common examples include austenitic stainless steels and certain aluminum alloys where dissimilar filler metals are chosen to reduce susceptibility to solidification cracking, although the mechanisms are different. Although the terms "homogenous" and "heterogeneous" technically differentiate the two types of fusion zones involving a filler metal, these terms are rarely used in practice.

Fusion zone microstructures vary greatly, and two examples are shown in Figure 9.4. The patterns in both the austenitic stainless steel and carbon steel microstructures shown provide evidence of dendritic solidification which is typical of most fusion zones. The solidification cracking associated with alloys such as austenitic stainless steels occurs along the boundaries between the dendrites, known as solidification grain boundaries. Solidification cracking is discussed in detail in Chapter 11. Carbon steel fusion zones are not susceptible to this type of cracking.

The unmixed zone (Figure 9.3) is an extremely narrow region of 100% melted base metal along the fusion line of a heterogenous fusion zone. Because it is so narrow, it is often difficult to distinguish. In some cases, the mechanical properties or corrosion properties of this zone can be degraded. Although it may be difficult to identify, an unmixed zone is theoretically present in every fusion weld. This is because although most of the weld puddle is somewhat turbulent, there must be a stagnant layer of melted base metal where the liquid metal meets the solid metal (fusion line). In most cases, however, the UMZ is not an important consideration.

9.3 The Partially Melted Zone

The partially melted (Figure 9.3) zone that occurs when welding some metals is a transition region between the fusion zone and the true HAZ. Many factors can contribute to forming partially melted zone, including segregation of alloying and impurity elements to the grain boundaries than can occur during base metal processing. These localized compositional variations

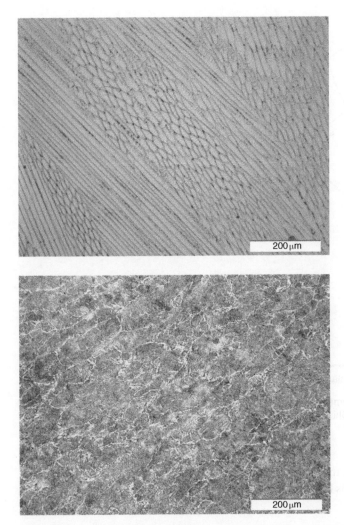

Figure 9.4 Fusion zone microstructures can vary greatly—an austenitic stainless steel fusion zone is shown at the top and a carbon steel fusion zone at the bottom.

may lower the melting temperature along the grain boundaries. As Figure 9.5 indicates, if the temperatures associated with the thermal gradient created by the welding process exceed the localized lower melting temperatures, an event known as grain boundary liquation may occur. The extent to which this occurs depends on many factors, including the slope of the thermal gradient and the amount of alloy and impurity segregation. Welding processes that produce greater heat input would be expected to create a wider partially melted zone due to a flatter thermal gradient.

Another phenomenon known as constitutional liquation can also cause liquated grain boundaries. In this case, when particles such as carbides begin to dissolve in the near HAZ, they may be the source of an element that diffuses into the surrounding matrix and effectively lowers its melting point. If the HAZ temperatures exceed the lower localized melting temperatures around the particle, a pool of liquid will form, and if the particle is located at a grain boundary, the liquid may flow along (or wet) the grain boundary resulting in a liquated grain boundary. Because liquated

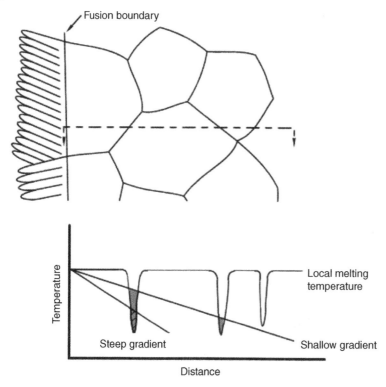

Figure 9.5 Segregation of alloying elements and impurities along grain boundaries can result in localized melting. (*Source:* Dr. John Lippold, The Ohio State University).

grain boundaries have no strength, they easily pull apart as the weld cools and shrinks creating stresses that form what are known as liquation cracks.

9.4 The Heat-Affected Zone (HAZ)

Although the weld HAZ often includes the partially melted region, the true HAZ (Figure 9.3) is the region between the unaffected base metal and the partially melted zone in which all metallurgical reactions are solid-state. Microstructure evolution in the true HAZ can be quite complex, depending on alloy composition, prior processing history, and welding heat input. A wide range of microstructures is possible within an HAZ, and materials that undergo phase transformations (such as steels) during heating and cooling tend to have the most distinct and complex HAZ microstructures (Figure 9.6). For example, steels transform to austenite (fcc) when heated above a critical temperature, and then back to ferrite (bcc) phase when cooled. But if cooling rates in the HAZ are fast enough, nonequilibrium phases such as martensite may form resulting in a microstructure that may be much different than that of the base metal. Adding further complexity to understanding an HAZ is the fact that both peak temperatures and cooling rates vary across the zone. Fastest cooling rates and highest peak temperatures occur closest to the fusion line. And whereas equilibrium binary phase diagrams can sometimes be used to predict HAZ microstructures, metal alloys that exhibit phase transformations such as steels may form weld

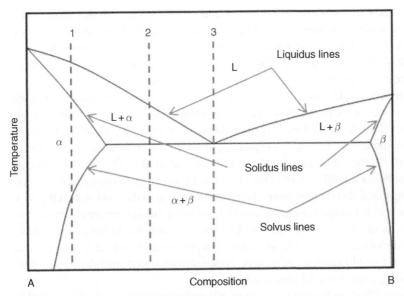

Figure 9.6 Simple binary eutectic equilibrium phase diagram with weld solidification and phase transformation sequence for three different alloys.

microstructures that are not predicted by equilibrium phase diagrams due to the nonequilibrium cooling conditions that are typical of welds. In this case, other diagrams which factor in cooling rates have been developed and will be covered in the next chapter on welding metallurgy of carbon steels.

As discussed earlier in this chapter, the weld HAZ is a function of the metal type and the base metal processing history. An aluminum alloy that was strengthened by strain hardening will be subject to softening in the HAZ, while some nickel alloys suffer from excessive strengthening mechanisms in the HAZ that can produce cracking. And HAZs that are formed when welding metals that do not undergo solid-state phase transformations during cooling (such as aluminum and nickel) are not affected by fast cooling rates in the way steel is. These subjects are also discussed in more detail in the following chapters.

Solid-state welds may or may not exhibit an HAZ. Processes that rely on the generation of a significant amount of heat such as Friction Welding and Resistance Flash Welding will produce an HAZ. But processes such as Explosion Welding, Ultrasonic Welding, and Diffusion Welding rely more on pressure and/or time than high heat, and therefore, do not produce an HAZ.

9.5 Introduction to Phase Diagrams

Approaches to studying and understanding welding metallurgy often rely on the use of phase diagrams. A phase diagram defines the equilibrium phases present in a metal alloy as a function of temperature and composition. Phase diagrams can be very complex, but the most common and simple type is the binary phase diagram that defines the stability of phases between two metals, or between one metal and one nonmetal. Binary phase diagrams can be used to predict and understand HAZ transformations as well as how a weld solidifies and the subsequent microstructures that form in both regions.

A simple binary eutectic phase diagram of element A and B (Figure 9.6) provides some examples. Phase diagrams always contain liquidus, solidus, and solvus lines. The liquidus lines separate all liquid material from the mixture of liquid and solid. The solidus lines separate completely solid material from the mixture of liquid and solid. And the solvus lines provide information regarding the extent to which one element can be completely dissolved in the other.

As indicated on the figure, a weld fusion zone of composition 1 would first pass through a liquid and solid phase as it cools from the liquid phase. This is sometimes referred to as a "mushy" stage since there is a mixture of both liquid and solid metal in this temperature range (or phase field). As it cools from the liquid, the line indicating the beginning of solidification is known as the "liquidus." Upon further cooling, the remaining liquid solidifies as indicated by the "solidus" line. At composition 1, all of the liquid solidifies as an α (alpha) phase. Upon further cooling, the α "solvus" line is passed, indicating that β (beta) phase would be expected to form in the solid-state if the cooling rates were slow enough. It is important to understand that the α phase represents element "A" with some dissolved element "B," and the α "solvus" line represents the maximum amount of element "B" that can be dissolved in metal "A" at a given temperature. In summary, a metal that solidifies as composition 1 would be expected to have a microstructure that solidifies as all α with some β phase precipitating later as a solid-state reaction.

At composition 2, solidification begins to occur at a lower temperature, and upon further cooling the horizontal line indicating the eutectic temperature is reached. At the eutectic temperature, all remaining liquid solidifies as the eutectic composition of $\alpha + \beta$. This is very different from composition 1 that solidified as 100% α. As a result, the weld fusion zone microstructure of composition 2 would be expected to be very different from that of composition 1. The composition 2 microstructure would consist of islands of primary α phase (representing the solidification up to the point of reaching the eutectic temperature) surrounded by the eutectic phase. The eutectic phase is a mixture of α and β, with the proportions of each dictated by the phase diagram.

Finally, composition 3 represents the eutectic composition. In this case, all the liquid immediately solidifies as the eutectic phase upon cooling to the eutectic temperature. This microstructure would consist of 100% eutectic phase, a mixture of α and β. In summary, the phase diagram predicts that the three different compositions of a mixture of element A and B would all be expected to produce significantly different weld fusion zone microstructures.

Actual binary phase diagrams can range from relatively simple to very complex (Figure 9.7). Three-dimensional ternary diagrams that describe the phase equilibrium of a mixture of three elements are even more complex. But again, these equilibrium diagrams should only be used for theoretical knowledge and predictive purposes since very few metal alloys contain only two elements, and weld solidification never occurs under the equilibrium conditions that the diagrams represent. Equilibrium conditions assume infinite time for diffusion of elements, something that never occurs when welding. Recently, as described in the previous chapter, powerful modeling software such as ThermoCalc™ is becoming a more common method for predicting weld microstructures. These software packages allow the user to generate phase diagrams based on multiple alloying elements and nonequilibrium solidification conditions to portray solidification of actual weld metals more accurately.

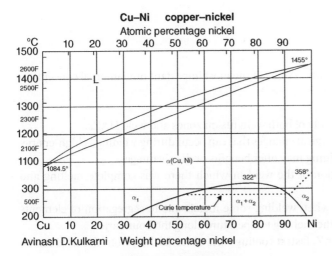

Cu–Ni   copper–nickel

Atomic percentage nickel

Avinash D.Kulkarni   Weight percentage nickel

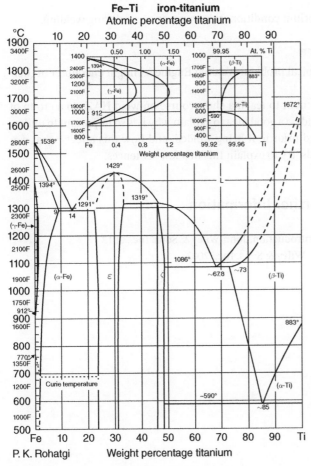

Fe–Ti   iron-titanium

Atomic percentage titanium

P. K. Rohatgi   Weight percentage titanium

Figure 9.7 Two examples of actual binary phase diagrams: the simple Cu-Ni system at the top, and the complex Fe-Ti system on the bottom. (*Source: ASM Metals Handbook*, Volume 8, Eighth Edition).

9.6 Test Your Knowledge

I) Fundamental Concepts—True/False

The following true/false questions pertain to some of the most important fundamental concepts in this chapter:

1) Welding metallurgy refers to the study of both weld fusion zones and weld HAZs.
2) When considering possible metallurgical changes that can occur during welding, a high-energy welding process such as Laser Welding may offer benefits or disadvantages.
3) The fusion zone refers to the region of the weld in which there was complete melting and solidification.
4) The partially melted zone occurs when welding some metals due to segregation of elements along grain boundaries that raise the melting temperature along the boundaries.
5) During the formation of a weld HAZ, fastest cooling rates occur farthest from the weld fusion line.
6) Phase diagrams are based on equilibrium conditions which never occur during welding.

II) Solve a Welding Engineering Problem

The following challenge represents a typical problem Welding Engineers might encounter in their career:

You have been given a unique assignment to teach a short course on weld HAZs to a group of welders. To introduce the course, it would be a good idea to explain why it is so important to understand how the base metal was processed prior to it being welded, and how this affects what can occur in the HAZ. Using your own words, try to explain this in a few sentences.

Recommended Reading for Further Information

Materials Science and Engineering—An Introduction, John Wiley & Sons, Inc., 2013.
Welding Metallurgy, Second Edition, John Wiley & Sons, Inc., 2003.

10

Welding Metallurgy of Carbon Steels

10.1 Introduction to Steels

Steels continue to be the most common commercially produced metal alloys. These alloys of iron and carbon, also referred to as ferrous alloys, account for over 90% of the metals produced and used on earth. Iron becomes steel when small amounts of carbon are added which produces a profound interstitial strengthening effect. Steels generally contain between 0.05 and 0.8% carbon by weight. Further strengthening is achieved through other alloying additions and phase transformation strengthening, which takes advantage of what is known as the allotropic behavior of iron. Steels can range from very simple iron alloys containing mainly carbon and manganese, to much more complex alloys with multiple alloying additions.

Structural steels can be generally classified into four groups based on composition—plain carbon steels, low alloy steels, high-strength low alloy (HSLA) steels, and high alloy steels. Plain carbon steels are simple Fe-C alloys with small additions of Mn and Si. They are sometimes distinguished by their relative carbon content—low carbon (<0.2%), medium carbon (0.2–0.4%), high carbon (0.4–1.0%), and ultra-high carbon (1.0–2.0%). Low alloy steels may contain up to 8% total alloy additions and may have low or medium carbon contents. Many of these steels are quenched and tempered to achieve high strength. High strength low alloy steels encompass a wide range of compositions. Typically, these steels have low carbon content, and achieve their strength through special processing techniques such as controlled rolling, or through micro-alloying additions that promote small grain size and/or precipitation reactions. The high alloy steels are used primarily at elevated temperatures where strength and corrosion resistance are important. Chromium is the alloying element normally added to impart the corrosion resistance. High alloy steels containing 12% chromium or greater are considered stainless steels (discussed in the next chapter).

A variety of classification systems are used for steel in the United States. The American Iron and Steel Institute (AISI)/Society of Automotive Engineers (SAE) is the most widely used system and is based on a four-digit classification system. The first digit indicates the predominant alloying element(s), while the second number reveals additional information about alloying elements or their amounts. The last two digits when divided by 100 reveal the amount of carbon in weight percent. Two common steels, 1018 and 4340, are used as examples below.

Example 1 (1018): The first digit which is a "1" indicates this is a carbon steel. The "0" which follows means this steel is a carbon steel with no additional alloying of note. The "18," after dividing by 100, describes steel that nominally contains 0.18 wt.% carbon.

Welding Engineering: An Introduction, Second Edition. David H. Phillips.
© 2023 John Wiley & Sons, Inc. Published 2023 by John Wiley & Sons, Inc.
Companion Website: www.wiley.com/go/Phillips/WeldingEngineeringIntroduction

> **Example 2 (4340):** In this case, the initial number "4" reveals that this is a molybdenum steel, and the "3" that follows indicates that chromium and nickel are also added. Per the comment above, the last two numbers indicate this steel contains 0.40 wt.% carbon.

The American Society for Testing and Materials (ASTM) specifies steels based on mechanical properties rather than on composition limits. The various ASTM specifications contain general requirements for broad families of steel products, such as bar, plate, sheet, and so on. Since they are specified by their mechanical properties, considerable latitude is typically allowed regarding their chemical composition. Steels under this system are identified by a letter followed by a random number which refers to the ASTM specification to which they are made. For example, A322 refers to hot rolled alloy steel bars. The letter "A" indicates a ferrous alloy, and the number "322" refers to the specification number for this steel product.

The American Society for Mechanical Engineers (ASME) Boiler and Pressure Vessel code uses a system known as "P-Numbers" which places metal alloys such as steels in broad groups based on their weldability characteristics. The letter P is followed by a number which refers to the weldability group the alloy falls under. For example, in designation P-Number 1, the "1" refers to a group of steels that all exhibit similar weldability characteristics. P-Numbers for ferrous alloys range between 1 and 11 (although the number 2 is not used). The advantage of this approach when welding a fabrication is that it allows for a change in the base metal alloy type without the need to requalify the welding procedure, as long as the change is to an alloy type that is in the same weldability group (P-Number).

Iron exhibits a somewhat unique property among commercial metals known as allotropic behavior (titanium is allotropic as well), which means it can exist as different crystal forms purely as a function of its temperature (Figure 10.1). At low temperatures, below 912°C (1674°F), iron has a body-centered cubic (bcc) crystal structure known as alpha ferrite. At intermediate temperatures, above 912°C and below 1394°C (2541°F), it forms a face-centered cubic (fcc) structure known as austenite, and at very high temperatures, above 1394°C and below the melting temperature, it again forms a bcc structure known as delta ferrite. The addition of carbon to produce steel has some effect on these transition temperatures. As will be discussed later, the allotropic behavior can be a benefit during processing but a detriment during welding.

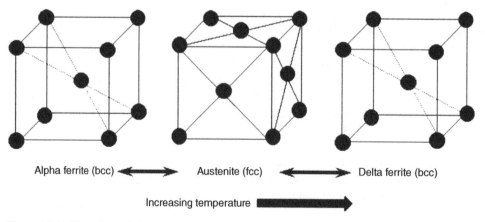

Alpha ferrite (bcc) ⟷ Austenite (fcc) ⟷ Delta ferrite (bcc)

Increasing temperature ⟹

Figure 10.1 The allotropic behavior of iron.

10.2   Steel Microstructures and the Iron-Iron Carbide Diagram

The physical metallurgy of steels is based on the well-known iron-iron carbide diagram shown in Figure 10.2, which denotes a small portion of the iron-carbon diagram. The right side of this diagram (6.67% C) represents the composition of an intermetallic phase with a stoichiometry of Fe_3C known as cementite or iron carbide. The diagram shown is not drawn to scale to better reveal the important phase fields in the iron-rich side of the diagram, which is where most steel compositions exist.

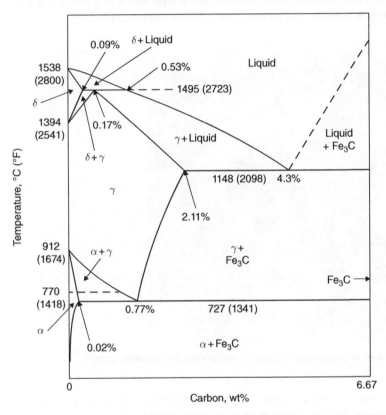

Figure 10.2   The iron-iron carbide phase diagram. (*Source:* Dr. John Lippold, The Ohio State University).

The three phases of iron can be seen on the left side of this diagram. The high temperature bcc delta (δ) ferrite is interesting but of minimal importance to steels. What is of particular importance to the processing of steels is the transition from the intermediate temperature fcc austenite (γ) to the lower temperature bcc alpha (α) ferrite. Notice that the austenite phase field extends far to the right meaning that austenite can dissolve a significant amount of carbon (up to 2.11%). On the other hand, the maximum amount of carbon that can be dissolved in the alpha ferrite is 0.02%. Since virtually all steels contain greater than 0.02% carbon, the consequence of this difference is that upon equilibrium cooling from austenite, any carbon in excess of 0.02% will exceed the solubility limit of ferrite and result in the additional formation of cementite (Fe_3C).

Depending on the rate of cooling from austenite, the Fe_3C cementite may take on various forms. If cooling rates are extremely slow allowing for significant diffusion, rounded particles of Fe_3C would be expected to form in a matrix of ferrite. This is the equilibrium morphology known as spheroidite. In practice, cooling rates during processing or welding are never slow enough to form spheroidite.

When cooling rates are fast enough to create nonequilibrium conditions, but are still relatively slow, the constituent that typically forms is a layered structure known as pearlite (Figure 10.3). The name pearlite reflects the fact that it often looks like mother-of-pearl when viewed under a microstructure. It consists of colonies of thin layers of Fe_3C (6.67% carbon) and ferrite (0.02% carbon) that form upon cooling from austenite. The layering morphology occurs because diffusion distances in a layered structure are minimal creating the "easiest" path for the excess carbon (above the maximum amount ferrite can dissolve, 0.02%) to diffuse out of the austenite into the high carbon Fe_3C cementite. Faster cooling rates from austenite temperatures may produce transformation products known as martensite and bainite, which will be discussed later.

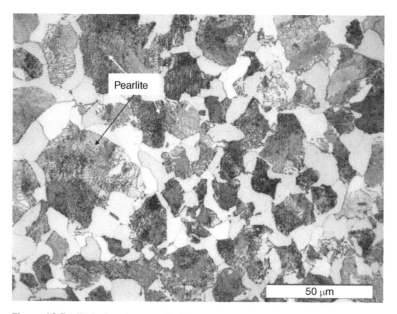

Figure 10.3 Typical carbon steel microstructure of pearlite and primary ferrite (white phase).

To reflect the fact that pearlite is such a common constituent that forms in steels upon cooling from austenite, a modification of the iron-iron carbide diagram (Figure 10.4) is often used. In addition to revealing constituents such as pearlite, this diagram reveals other important features associated with the processing and welding of steels. The A_3 and A_1 temperatures represent important temperatures as the steel cools from austenite, such as during a heat treatment or when it is being welded. At the A_3 temperature, ferrite will begin to form from the austenite (at compositions below the eutectoid composition). Upon further cooling to the A_1 temperature, all remaining austenite will transform to pearlite. Again, this is the case for relatively slow cooling rates.

The A_1 temperature, 727°C (1341°F), is also known as the eutectoid temperature, another important feature in this equilibrium phase diagram. A eutectoid reaction is similar to a eutectic reaction during which a single liquid phase transforms into a two-phase solid at a specific temperature. The difference with the eutectoid reaction of course is that there is no liquid involved; a single phase solid (austenite) transforms to a two-phase solid (ferrite + Fe_3C, or pearlite). Under relatively slow cooling conditions, the microstructure of a carbon steel containing less than the eutectoid composition (0.77%) of carbon would consist of a microstructure of alpha phase (or ferrite) that begins to form at the A_3 temperature, surrounded by pearlite that forms from the remaining austenite upon

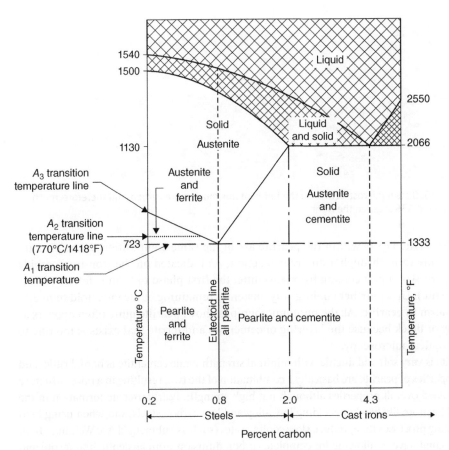

Figure 10.4 Modification and simplification of the iron-iron carbide diagram. (*Source: Welding Essentials,* Second Edition).

reaching the A_1 temperature. At the eutectoid composition of 0.77%, all the austenite transforms directly to pearlite resulting in a microstructure of 100% pearlite. The A_2 temperature is known as the curie temperature, above which iron is no longer ferromagnetic.

As the diagram indicates, steels that contain greater than the eutectoid composition of carbon would first form Fe_3C cementite upon cooling from austenite. Upon reaching the eutectoid temperature, all remaining austenite would transform to pearlite. The result is a microstructure of primary cementite and pearlite. These very high carbon steels are known as "hypereutectoid" steels, while those with compositions below the eutectoid composition are known as "hypoeutectoid." This diagram also reveals that cast irons consist of iron with extremely high amounts of carbon, in excess of 2%.

Most steels are "hypoeutectoid" steels. Again, assuming relatively slow cooling rates from austenite temperatures, these steels will exhibit a microstructure of ferrite and pearlite. Since pearlite contains the high carbon cementite, higher carbon "hypoeutectoid" steels will consist of a higher ratio of pearlite to ferrite. Figure 10.5 compares the microstructures of a relatively high carbon hypoeutectoid steel (1060) on the left to a much lower carbon (A36) steel.

The white phase in these microstructures is known as primary ferrite (the word primary distinguishes it from the layers of ferrite in the pearlite), while the dark gray/black constituent is the pearlite. As shown previously in Figure 10.3, the pearlite consists of layers of cementite

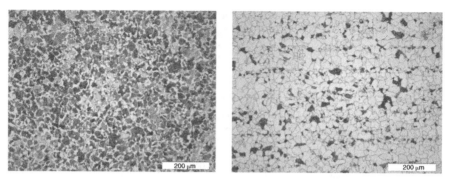

Figure 10.5 The 1060 steel microstructure on the left contains much more carbon, and therefore, much more pearlite than the A36 steel on the right.

and ferrite. The ferrite layers within the pearlite are of the same composition as the primary ferrite, they simply form through a different reaction. As indicated on the iron-iron carbide diagram of Figure 10.4, upon cooling from austenite, the first phase to form is ferrite (this is the primary ferrite). Upon further cooling, any austenite remaining at the eutectoid temperature will transform to pearlite. As the photomicrographs show, pearlite quite often appears as solid dark gray or black because the layering of cementite and ferrite that exists is too fine to resolve under optical microscopy.

Primary ferrite is very soft and ductile with minimal strength while cementite is hard, brittle, and strong. The properties of pearlite are basically a combination of the two, resulting in a microstructure with generally good overall properties although not high strength. Pearlite/ferrite formation in the weld HAZ will be more likely when welding low alloy and low carbon steels, and when using high heat input welding processes that produce slow cooling rates (such as Submerged Arc Welding). Both cases are more conducive to allowing for complete carbon diffusion from austenite into ferrite and pearlite, and for the fcc austenite atoms to rearrange to form bcc ferrite and pearlite (Figure 10.6).

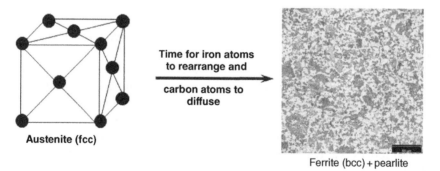

Figure 10.6 When cooling rates from austenite are sufficiently slow, a microstructure of ferrite + pearlite can be expected. (*Source:* Wayne Papageorge, The Ohio State University).

However, for a given steel, when cooling rates are relatively fast there may not be enough time for the fcc austenite atoms to rearrange to form bcc ferrite, and for the carbon to diffuse out of the austenite to form ferrite and pearlite. This may result in the formation of a diffusionless body centered tetragonal (BCT) structure known as martensite (Figure 10.7). The BCT crystal structure is essentially a "stretched" bcc structure that results from a shearing mechanism due to the excessive

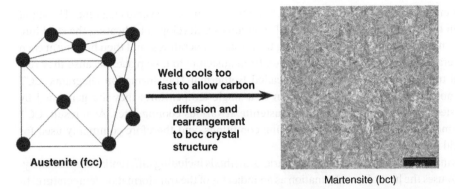

Weld cools too
fast to allow carbon

diffusion and
rearrangement
to bcc crystal
structure

Austenite (fcc)

Martensite (bct)

Figure 10.7 When cooling rates from austenite are sufficiently fast, a microstructure of martensite can be expected. (*Source:* Wayne Papageorge, The Ohio State University).

carbon (above the 0.02% that the bcc ferrite can dissolve) being trapped and extending the structure in one direction. When viewed under an optical microscope, martensite typically exhibits a needle-like morphology. It is known to be hard and brittle, with hardness levels that increase with increasing carbon content. Hard martensite is known to be susceptible to a potentially catastrophic cracking mechanism known as hydrogen cracking.

To recover some ductility, martensite is typically given a relatively low temperature heat treatment known as tempering. Tempering softens the martensite through the formation of fine carbides, which reduces the amount of trapped carbon in the BCT matrix. Depending on the tempering time and temperature, some ferrite formation may occur as well, which also contributes to the softening. Martensite formation will be more likely to form in the weld HAZ when welding high alloy steels with higher amounts of carbon (such as 4340) and low heat input processes that create fast cooling rates (such as Laser Welding).

One more important transformation product that may form when cooling from austenite is bainite, which is typically a fine needle-like structure consisting of ferrite and cementite. Because its features are so fine, it is often difficult to identify with conventional optical microscopy. In contrast to the formation of martensite, bainite formation involves some diffusion. Because bainite offers higher strength than ferrite + pearlite, and better ductility than martensite, it has become a desirable phase in the processing of many modern steels. For example, quenched-and-tempered steels often contain a mixture of martensite and bainite. A typical heat treatment to form bainite involves cooling fast enough to avoid transformation to pearlite/ferrite, and then holding at an intermediate temperature for some time.

10.3 Continuous Cooling Transformation (CCT) Diagrams

So, while the iron-iron carbide phase diagram can be used to determine the phase balance in steels under equilibrium conditions of very slow cooling, such conditions are usually not the case during welding where rapid cooling rates occur. As a result, another type of diagram, which considers cooling rates, is needed. Such diagrams are known as Time Temperature Transformation (TTT) and Continuous Cooling Transformation (CCT) diagrams. TTT and CCT diagrams can be used to predict steel microstructures as a function of cooling rate from austenite temperatures. Each diagram is only applicable to a single steel composition.

These diagrams are similar in that they are both plots of temperature versus log time. The slight difference is in the way they are generated. TTT diagrams are developed by heating the steel into the austenite temperature range, rapidly cooling to various temperatures, and then holding at each of these temperatures to allow for transformation from austenite to take place. Because they rely on isothermal transformation, they are often called Isothermal Transformation Diagrams. CCT diagrams do not involve holding the specimen at a single temperature; they are generated by allowing the steel to cool continuously from austenite at various cooling rates. As a result, CCT diagrams are more representative of real welding conditions, and therefore commonly used for predicting weld microstructures.

The transformation timing is determined by various methods including differential thermal analysis (DTA). DTA uses the heat of transformation as an indicator of the transformation temperature. In steels, considerable latent heat is liberated when austenite transforms to ferrite, pearlite, and other transformation products. This heat is detected through the DTA measurement technique resulting in individual points that can be plotted on the diagram. Each isothermal hold (TTT diagrams) or cooling rate (CCT diagrams) will generate a unique set of points. By repeating this step at different temperatures/cooling rates, a set of transformation curves can be generated, which indicate both the start and completion of the transformation. Figure 10.8 is a very simple example of a TTT diagram superimposed with various cooling rates. Since this diagram only shows one transformation start curve and one transformation end curve, it has to represent a steel of eutectoid composition (0.77% carbon). Such a steel can only form 100% pearlite (no ferrite) upon cooling from austenite, assuming the cooling rates are not fast enough to form martensite. At compositions less than the eutectoid composition (typical of most steels), an additional ferrite transformation curve will be present.

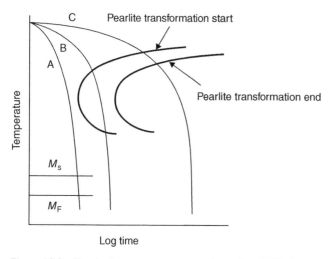

Figure 10.8 Simple time temperature transformation (TTT) diagram with various cooling rates superimposed.

TTT and CCT diagrams typically consist of ferrite and pearlite transformation curves, and martensite start (M_s) and martensite finish (M_f) temperatures. The Ms denotes the temperature at which martensite formation begins, and the M_f is the temperature at which martensite transformation is complete. The transformation curves represent the beginning and completion of ferrite/pearlite. Once the ferrite/pearlite transformation from austenite is complete, no other microstructures (such as martensite) can form. Or described another way, ferrite, pearlite, and martensite can only form from austenite.

The diagram shown also includes cooling rates that might be expected during a typical heat treatment process or in a weld HAZ. In this case, a cooling rate as fast as curve "A" would produce 100% martensite. By avoiding the nose of the curve, all austenite transforms to martensite. Cooling curve "B" would produce a pearlite + martensite microstructure because the pearlite transformation is not completed, meaning there is still some austenite left to be transformed to martensite upon reaching the martensite start temperature. A cooling rate such as "C" would be expected to form 100% pearlite since the pearlite transformation is completed. As mentioned above, once all austenite has been transformed to pearlite, it is no longer possible for any other transformation products to form. It is also important to point out that the TTT and CCT diagrams are always based on cooling from temperatures in the austenite phase field. These diagrams cannot be used to predict what happens upon heating, only what happens upon cooling from austenite. Finally, it's important to mention that the cooling rate between 800°C (1472°F) and 500°C (932°F) is critical since this is the temperature range where the transformations from austenite take place.

The diagram of Figure 10.8 is relatively simple since it only involves pearlite transformation curves. In most cases, these diagrams are more complex as indicated in Figure 10.9, a CCT diagram for a 1040 steel. This diagram reveals that a wide range of microstructures that are possible as a function of the cooling rate from austenite temperatures, even with a steel as simple as 1040.

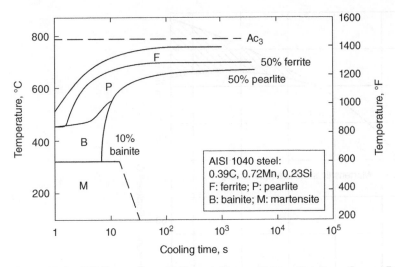

Figure 10.9 CCT diagram for a 1040 steel. (*Source: Welding Metallurgy*, Second Edition, Figure 17.10, page 404. Reproduced with permission from John Wiley & Sons, Inc.).

Martensite is a very important microstructural constituent in steels with both positive and negative connotations. Because of its high strength, many steels such as 4340 rely on its formation during processing. This form of strengthening is known as phase transformation strengthening since it is associated with the transformation from austenite to martensite. As mentioned previously, a subsequent tempering heat treatment allows the steel manufacturer to customize properties to create a balance of strength and ductility. Longer temper times and/or higher temper temperatures will produce more ductility, but at the expense of strength. Steels that are processed this way are sometimes referred to as "quenched and tempered" steels.

Quenched and tempered steels are designed to form martensite quite easily. This is primarily achieved through the addition of alloying elements such as chromium, nickel, and molybdenum. Martensite can become problematic when these types of steels are welded. Weld fusion zone and

HAZ martensite can form quite easily, which in its untempered state is very hard with poor toughness and poor ductility, and is susceptible to hydrogen cracking.

Steels that form hard martensite during welding almost always require a postweld temper treatment, which in some cases may be time consuming and expensive. For example, while it may be relatively easy to temper a small plate by placing it in a furnace, larger fabrications such as pipelines or large assemblies may require the use of localized heating blankets or resistance heaters. Difficulties in conducting a tempering treatment may occur due to a variety of field implementation issues, such as intricate geometry and access issues.

10.4 Hardness and Hardenability

As shown in Figure 10.10, the hardness of martensite is directly related to the carbon content of the steel. Martensite with higher hardness will be more susceptible to hydrogen cracking. In cases of high restraint and high hardness, it may be difficult to weld without cracking, so quite often preheating and/or weld interpass temperature control is used to keep cooling rates low and allow

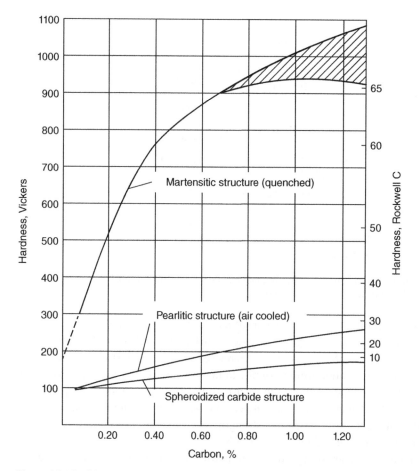

Figure 10.10 Plot of hardness versus carbon content for martensitic, pearlitic, and spheroidized microstructures.

hydrogen to diffuse out of the weld region. The figure also shows that the effect of carbon content on the hardness of pearlitic microstructures is much less profound.

When studying the processing and welding of steels it is important to understand the concepts of hardness and hardenability. Hardenability is one of the most important concepts pertaining to structural steels and is most simply defined as the ease with which a given steel forms martensite. Maximum hardness is achieved when a fully martensitic structure is formed, so steels that can be fully hardened over wide ranges of cooling rates are deemed to have good hardenability. The CCT diagram can provide immediate evidence of a steel's hardenability. CCT curves that are shifted more to the right (longer times) indicate greater hardenability since the formation of 100% martensite can occur at slower cooling rates as compared to a less hardenable steel.

There are several factors that control the hardenability of a steel. Alloying additions have the greatest effect on hardenability because they are effective in delaying the transformation to other transformation products (ferrite, pearlite, bainite) during cooling. Large grain sizes will slow the transformation to ferrite/pearlite allowing martensite to form, thereby increasing hardenability. The coarse-grained HAZ of steels have large grain sizes so martensite may more easily form in this region than what the CCT diagram predicts. Increases in carbon content not only increase the hardness of the martensite but also play a role in increasing the hardenability as well.

Finally, the thickness of the steel indirectly plays a role in hardenability since it affects cooling rates. During heat treating, thin plates or small diameter bars may be readily hardened by quenching since heat extraction is quite rapid. Hardening thick plates or large forgings this way is more challenging because of the variation in cooling rate from surface to center. But when welding, thicker sections will produce faster weld cooling rates that increase the likelihood for martensite formation. This is because with thicker plates, cooling occurs in three dimensions instead of two (Figure 10.11) as discussed previously.

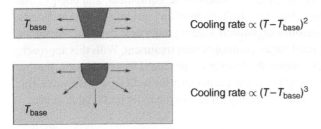

Figure 10.11 Due to 3-D heat flow, weld cooling rates are greater when welding thicker cross-sections.

Another way to assess and compare hardenabilities of different steels is through a formula known as "carbon equivalent" (CE). The "carbon equivalent" combines carbon with other alloying additions in a formula that can be used to quantitatively predict the hardenability of the steel. There are many versions of this formula, but one of the most common is known as the IIW (International Institute for Welding) CE (Equation 10.1):

$$CE_{IIW} = \%C + \%MN / 6 + \%(CR + Mo + V) / 5 + \%(Si + Ni + Cu) / 15 \qquad (10.1)$$

The higher the CE, the greater the hardenability of the steel. The primary purpose of the carbon equivalent formula is to determine the relative probability of forming hard martensite during welding, and thereby, predict the relative susceptibility to hydrogen cracking. To avoid hydrogen cracking, the CE calculation can be used as a guideline for making determinations about preheating and postweld heat treating. These determinations are affected by many factors including hydrogen content and residual stress, but the following guideline is a typical example:

- CE < 0.35   No preheat or postweld heat treatment
- 0.35 < CE < 0.55   Preheat
- 0.55 < CE   Preheat and postweld heat treatment

As discussed, preheating slows the cooling rates to avoid forming 100% martensite. This is because after preheating, the surrounding base metal temperature is closer to weld zone temperatures so heat flow out of the weld zone is reduced (refer back to the formulas for cooling rates). Preheating can also help reduce residual stresses and distortion and reduce hydrogen content by promoting hydrogen diffusion out of the weld region. CCT diagrams can be used to estimate the desired cooling rates achieved through preheating to avoid martensite. For the reasons just discussed, when welding thicker plates or fabrications, higher preheat temperatures will be required to achieve the desired cooling rates. Typical preheat temperatures of quenched and tempered steels are in the neighborhood of 250°C (482°F).

In a manufacturing or fabrication operation, time, equipment, and energy costs associated with preheating may greatly increase the cost of welding. Also, in confined spaces or warm locations, high preheat temperatures may become a major source of discomfort for the welder, and larger fabrications may be very difficult to preheat. Nonetheless, preheating is often a requirement for many steels. Lack of a proper preheat may result in a catastrophic failure, or at least the need to scrap and/or repair the part or fabrication.

As mentioned, if martensite is produced during welding, its poor mechanical properties can be remedied through a postweld tempering heat treatment. Postweld heat treatments also help reduce any residual stress created during welding. Postweld tempering heat treatments are typically in the range of 400–600°C (750–1100°F). As with the preheat process, time, equipment, and energy costs associated with postweld heat treatments reduce the productivity of any welding operation. During multipass welding, "temper bead" or controlled deposition welding sequences have been designed that are self-tempering, eliminating the need for an additional heat treatment. With this approach, the heat from subsequent welding passes tempers the martensite produced by prior passes. Temper bead welding has been used successfully for many years, particularly for weld repair. However, this approach requires special welder training and is typically limited to the more hardenable steels.

10.5 Hydrogen Cracking

Hydrogen cracking (Figure 10.12) can occur in the weld HAZ or fusion zone, and typically occurs in regions of high stress concentration such as the weld toe, root, or under the weld (referred to as underbead cracking) where weld residual stresses can be high. For hydrogen cracking to occur, four conditions are required:

1) Susceptible microstructure
   - Martensite of high hardness
2) Source of hydrogen
   - Moisture in the flux
   - Condensation
   - Grease/oil on parts

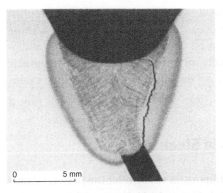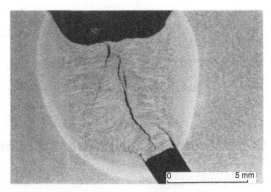

Figure 10.12 Hydrogen cracks from a Y-groove test—in the HAZ (left) and fusion zone (right). (*Source:* Dr. John Lippold, The Ohio State University).

3) Significant levels of tensile stress
   - Residual
   - Applied
4) Temperature range
   - Hydrogen cracking occurs between −100 and 200°C (−150 and 390°F)

Even if only one of the four conditions is eliminated, hydrogen cracking can be prevented. The most common approaches to avoiding hydrogen cracking emphasize eliminating the susceptible microstructure (hard martensite) using methods discussed previously, and reducing the level of hydrogen. Codes such as AWS D1.1 provide preheat, interpass temperature control (for multipass welding), and postweld heat treatment guidelines for controlling hard martensitic microstructures and avoiding hydrogen cracking.

When there is a concern for hydrogen cracking, it is also common to use what is known as "low hydrogen practice" when welding. Major sources of hydrogen include moisture in the flux with processes such as Shielded Metal, Flux Cored, and Submerged Arc Welding. Therefore, it is important to ensure that electrodes and fluxes are stored properly and are sufficiently dry prior to use. Processes that do not rely on a flux are less likely to be susceptible to hydrogen cracking, but moisture in the shielding gas used for the Gas Metal Arc and Gas Tungsten Arc Welding processes can provide a source of hydrogen and increase the risk. Aside from the welding consumables, paint, mill oil, degreasing fluids, condensed moisture, and heavy oxide layers on the part being welded also represent potential sources of hydrogen.

Tensile stress at the joint is one of the conditions that contribute to hydrogen cracking, but to rely on the elimination of such residual (or applied) stresses to control hydrogen cracking is usually not practical. However, if these stresses are extremely high, the chances of hydrogen cracking is increased, even if other hydrogen cracking control methods are being followed. Therefore, whenever there is a concern for hydrogen cracking, it is always a good idea to keep residual and applied stresses as low as possible. This can be accomplished by proper joint design and fixturing, control of weld bead size and shape, and weld bead sequence patterns in multipass welds.

Hydrogen cracking occurs at temperatures between −100 and 200°C (−150 and 400°F) and may not occur for hours or days after the initial weld is made, which is why it is sometimes referred to as delayed cracking. This is because hydrogen can easily diffuse to regions of high stress concentration and cause cracking well after the weld has cooled to room temperature.

Since it can occur well after an apparently crack-free weld is produced, it can be catastrophic, which is the reason why many carbon steel welding procedures require a 24–48 hour waiting period prior to inspection. When cracks occur, they can be detected by ultrasonic and radiographic methods; surface cracks may be detected visually, or with dye penetrant or magnetic particle inspection methods. Small cracked areas may be cut out and repair welded, while extensive cracking may result in scrapped parts.

10.6   Heat-Affected Zone Microstructures in Steel

When predicting the HAZ microstructure of steel, it is important to consider the peak temperatures that were reached in any given region of the HAZ. Figure 10.13 is a useful diagram for understanding the importance of peak temperature variation in the HAZ. Region 1, which is the portion of the HAZ that is closest to the fusion zone (known as the near HAZ), will experience temperatures high in the austenite phase field. The largest grain sizes form here, effectively increasing the likelihood for martensite to form. This is because as grain sizes increase, carbon diffusion is suppressed due to less grain boundary area. As a result, it takes more time for the transformation from austenite to ferrite/pearlite to occur, meaning there is a better chance that austenite will remain when the martensite start temperature is reached. Therefore, in the large grained HAZ, the nose of the CCT diagram for the particular steel being welded is effectively pushed to the right, indicating an increase in the likelihood of martensite formation here. This region is commonly known as the coarse-grained HAZ (CGHAZ) and is often the region most susceptible to hydrogen cracking. Farther out in the HAZ, the grain sizes become

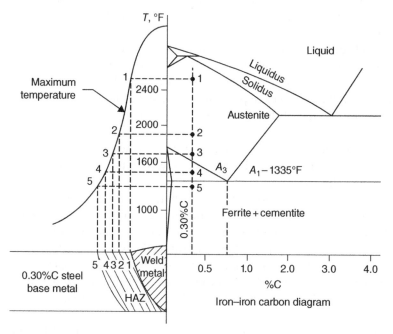

Figure 10.13   Typical variation of peak temperatures in the HAZ of a weld. (*Source: Welding Essentials, Second Edition*).

smaller due to temperatures reaching lower into the austenite field. This region (region 3) is known as the fine-grained HAZ (FGHAZ).

Even farther out in the HAZ (region 4) is a portion of the HAZ heated into the austenite plus ferrite phase field. This portion of the HAZ may contain a variety of microstructures, including a potential mixture of ferrite, pearlite, and martensite. For example, when heated just above the eutectoid temperature (A_1 temperature), any pearlite that was present in the base metal microstructure will begin to transform to austenite. Upon subsequent cooling, this austenite may transform back to a fine pearlite, or to martensite (or possibly bainite). This complex region is often referred to as the intercritical HAZ (ICHAZ). The ICHAZ can sometimes experience a significant loss in toughness because new austenite that forms from previous regions of pearlite will be high in carbon, and any subsequent martensite that forms will be very hard. If the steel being welded is a quenched and tempered steel, region 5 might consist of soft over-tempered martensite because this region effectively received an additional temper heat treatment during welding (in addition to the base metal processing temper).

Figure 10.14 shows a typical carbon steel weld microstructure. At the top of this photomicrograph is a portion of the fusion zone exhibiting evidence of dendritic solidification in the vertical direction. Immediately adjacent to the fusion zone is the near HAZ, clearly revealing the very large grain sizes typical of the coarse-grained HAZ region. As discussed above, the HAZ grain sizes get smaller at increasing distances from the fusion line, and at the very outer edge of the HAZ is the intercritical zone.

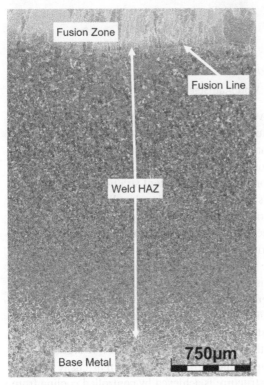

Figure 10.14 Typical carbon steel weld microstructure. (*Source:* Wayne Papageorge, The Ohio State University).

10.7 Advanced High-Strength Steels

The modern automobile is constructed of a wide array of a new generation steels known as Advanced High-Strength Steels. Steel manufacturers have developed numerous new steel processing techniques focused on generating high strength while maintaining good ductility (Figure 10.15). As compared to conventional steels, these steels contain higher alloying additions with complex multiphase microstructures, and generally rely on some combination of thermomechanical processing in the austenite phase field, and controlled cooling. They may also be further strengthened by strain-hardening. They are categorized by both strength (in MPa) and processing method. Tensile strengths range from 500 to over 2000 MPa, and ductility (% elongation) ranges from 10 to as high as 60%. For proprietary reasons, compositions are not always published.

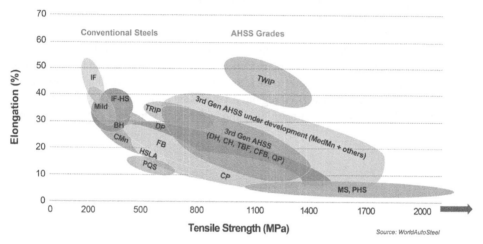

Figure 10.15 Advanced high-strength steels. (*Source:* WorldAutoSteel).

The development of these steels has occurred over what are referred to as three generations. The first generation are mostly ferrite-based with mixtures of some martensite and/or bainite, with possibly some retained austenite. They are known mainly for better ductility over conventional high-strength steels at the same strength level. The second generation of Advanced High-Strength Steels are a mostly austenitic microstructure that rely on a unique and potent strain-hardening mechanism known as mechanical twinning. These steels can have outstanding ductility due to the austenitic microstructure, while maintaining very good strengths. The third generation achieves the highest strengths of all these steels with microstructures that are mostly martensitic or bainitic with some retained austenite, and possibly some ferrite and/or carbides. As mentioned, these steels achieve such excellent properties through a wide variety of processing methods, all of which rely heavily on controlled cooling techniques from austenite. For example, Figure 10.16 demonstrates the different cooling paths from austenite for several common Advanced High-Strength Steels. It is beyond the scope of this book to describe in detail the processing methods, microstructures, and properties of all these steels. But to get an understanding of at least some of the approaches used, these three common families of steels will be briefly reviewed—DP, TRIP, and TWIP steels.

The Dual Phase (DP) steels consist of a two-phase microstructure of hard martensite and soft ferrite. As shown in the above figure, this microstructure is achieved by controlled cooling from austenite partially into the ferrite phase field to form some ferrite, followed by a rapid quench to transform the remaining austenite to martensite. The strength/ductility ratio can be controlled by changing the balance of martensite and ferrite. The highest strength (approaching 1000 Mpa)

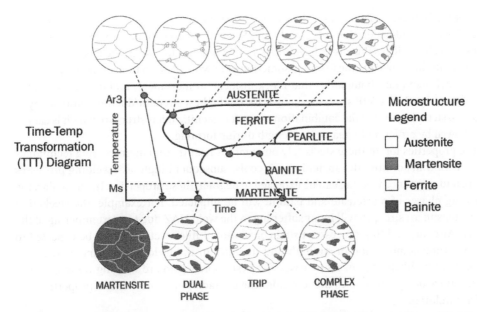

Figure 10.16 Controlled cooling paths for common advanced high-strength steels. (*Source:* WorldAutoSteel).

versions may contain greater than 50% martensite. Further strengthening of these steels can be accomplished during forming which work-hardens the ferrite, and through bake-hardening through the formation of carbides during curing in the paint bake ovens.

Transformation-Induced Plasticity (TRIP) steels are made up of a microstructure of mostly soft ferrite, with smaller amounts of hard martensite and bainite, and a minimum of 5% (by volume) retained austenite. As shown in Figure 10.16, the bainite is formed by an additional isothermal hold processing step in the bainite phase field. A microstructure and schematic of a typical TRIP microstructure is shown in Figure 10.17. The most unique and beneficial aspect of these steels is the excellent strain-hardening that can be achieved through strain-induced transformation of the austenite to martensite. Straining of the austenite during forming causes it to transform to high-strength martensite, and the extent of transformation is a function of the carbon content of the steel. Higher amounts of carbon stabilize the austenite which means greater amounts of strain will be needed to transform it to martensite. These attributes mean TRIP steels can be easy to form into automotive components as well as perform admirably during crash testing.

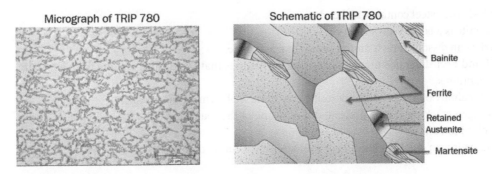

Figure 10.17 TRIP 780 microstructure and schematic. (*Source:* WorldAutoSteel).

The twinning-induced plasticity (TWIP) steels consist entirely of what is known as twinned austenite. These second-generation Advanced High-Strength Steels contain very high amounts of manganese which stabilizes the austenite down to room temperature. When the austenitic microstructure is strain-hardened during forming, something known as deformation twins are produced which act much like grain boundaries that block dislocation movement, rapidly increasing the strength of the steel. The TWIP steels are known for the highest strength/ductility ratio of any of these modern steels because of the combination of a soft austenitic microstructure which is easy to deform followed by profound increases in strength during forming.

Since the properties for all these steels rely on complex heat treatment processing techniques associated with transforming the austenite as it cools, subsequent high heat welding processes such as arc and resistance welding can be expected to degrade these properties. In particular, the strengthening phases of martensite and bainite will be impacted. For example, the steels that rely on martensite for strength will incur softening in the weld HAZ due to overtempering of the martensite. Also, arc welding processes using conventional filler metals cannot be expected to produce the same combination of strength and ductility in the weld fusion zone as the base metal. So, it seems likely that in many cases the best approach to arc welding these complex steels is to utilize design principles that consider the degradation of mechanical properties as a result of arc welding.

On the other hand, high weld hardness may be a problem. Since the vast majority of applications for these steels is in the automotive industry, Resistance Spot Welding tends to be the dominant welding process. This process produces rapid weld cooling rates due to the electrode cooling effect. Such cooling rates in combination with the high hardenability of these steels means high hardness can be expected in and around the weld nugget, potentially increasing the likelihood for hydrogen cracking. One approach to mitigating the high hardness is to incorporate an additional lower current weld pulse at the cycle to temper the hard martensite. Another major challenge associated with Resistance Spot Welding of these steels in the automotive industry is contending with a coating added for corrosion resistance known as galvanizing. These coatings produce extreme electrode wear, and in some cases, a weldability problem known as liquid metal embrittlement. However, a detailed discussion on these subjects is beyond the scope of this book.

10.8 Test Your Knowledge

I) Fundamental Concepts—True/False

The following true/false questions pertain to some of the most important fundamental concepts in this chapter:

1) An SAE 1018 steel contains 18% carbon.
2) Austenite is a bcc crystal structure.
3) Ferrite can dissolve much more carbon than austenite.
4) TTT and CCT diagrams predict phase transformations that occur upon cooling from austenite temperatures.
5) Hardenability of a steel can be described as the ease of forming martensite.
6) Hydrogen cracking can be prevented if only one of the four conditions for hydrogen cracking are eliminated.
7) Weld hydrogen cracking may occur well after the weld is completed.

8) The most likely location in a steel weld HAZ for martensite formation is along the very outer edge of the HAZ.
9) A steel weld HAZ may also contain a region of softening due to tempering of a previous martensitic base metal.
10) Advanced High-Strength steel processing relies heavily on controlled cooling techniques from austenite temperatures.
11) Advanced High-Strength steels exhibit extremely high strengths, but they have very poor ductility.

II) Solve a Welding Engineering Problem

The following challenge represents a typical problem Welding Engineers might encounter in their career:

The company you are working for is suddenly experiencing major weld cracking problems on a particular pressure vessel they manufacture. You are asked to investigate the problem, and you discover that the cracking began following a change to a newer, more highly alloyed steel. Also, the welders informed you that the cracks are not there immediately after welding but formed many hours later. Your Welding Engineering training now causes you to do a carbon equivalent (CE) calculation for this new steel which you determine to be 0.60. You now know what the problem is and how to solve it. How will you explain this to your boss?

Recommended Reading for Further Information

ASM Handbook, Tenth Edition, Volume 6—"Welding, Brazing, and Soldering," ASM International, 1993.
AWS Welding Handbook, Ninth Edition, Volume 4—"Materials and Applications, Part 1," American Welding Society, 2011.
Welding Metallurgy, Second Edition, John Wiley & Sons, Inc., 2003.

11

Welding Metallurgy of Stainless Steels

11.1 Introduction to Stainless Steels

Stainless steels represent a broad family of iron-based alloys that contain a minimum of approximately 12 wt.% chromium. The addition of chromium produces an extremely thin but stable and continuous chromium-rich oxide film that gives these alloys their stainless and corrosion-resistant properties. These steels are based on the iron-chromium, iron-chromium-carbon, and iron-chromium-nickel systems, but may contain other alloying additions that alter their microstructures or properties. In addition to resistance to discoloration and corrosion, they offer high-temperature oxidation resistance and a wide range of strength and ductility, depending on the alloy. Stainless steels are used for a wide variety of applications and in many environments. Applications vary from power generation and pulp-and-paper industries to common household products such as washing machines and kitchen sinks.

Stainless steels are generally grouped into five distinct alloy families and are primarily classified by the phase that dominates the microstructure. The five families of stainless steels are martensitic, ferritic, austenitic, duplex, and precipitation-hardening (known as "PH" stainless steels). Duplex stainless steels consist of a nearly equal mixture of ferrite and austenite, while the PH stainless steels can be either martensitic or austenitic. The American Iron and Steel Institute (AISI) uses a system based on three numbers, sometimes followed by a letter, to designate most stainless steels. Common examples are 304, 304L, 410, and 430. The duplex and PH stainless steels are designated differently, as will be discussed later in the chapter.

The iron-chromium phase diagram (Figure 11.1) forms the basis for stainless steels since Cr is the primary alloying element. Note that there is complete solubility of Cr in iron at elevated temperatures, and solidification of all Fe-Cr alloys occurs as ferrite. At low chromium concentrations a "loop" of austenite exists, commonly referred to as the "gamma loop." Alloys with greater than about 13% Cr will be fully ferritic at elevated temperatures, while those with less than this amount of Cr will form austenite within the gamma loop. Upon cooling, this austenite can transform to martensite.

Carbon is a potent austenite stabilizer which has the effect of expanding the gamma loop shown in the iron-chromium diagram. The gamma loop represents an important distinction between ferritic and martensitic stainless steels. Ferritic stainless steels contain low amounts of carbon which shrinks the gamma loop, thereby resulting in a mostly ferritic microstructure. Martensitic stainless steels, on the other hand, contain higher amounts of carbon, which expand the gamma loop allowing for the formation of austenite, which then easily transforms to martensite upon cooling.

Welding Engineering: An Introduction, Second Edition. David H. Phillips.
© 2023 John Wiley & Sons, Inc. Published 2023 by John Wiley & Sons, Inc.
Companion Website: www.wiley.com/go/Phillips/WeldingEngineeringIntroduction

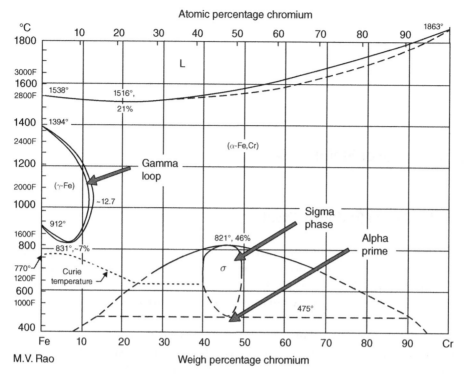

Figure 11.1 Fe-Cr binary phase diagram—the basis for all stainless steels. (*Source: Welding Metallurgy and Weldability of Stainless Steels*, Figure 2.1, page 9. Reproduced with permission from John Wiley & Sons).

As indicated in the phase diagram, a low temperature equilibrium phase, called the sigma phase, also forms in the Fe-Cr system. This phase has a FeCr stoichiometry, and thus is most likely to form in high Cr alloys. The kinetics of sigma phase formation is generally quite sluggish and requires extended time in the temperature range from 500 to 700°C (930 to 1300°F). Because sigma phase is hard and brittle, its presence in stainless steels is usually undesirable. Since its formation takes time, it is typically not a problem associated with welding processes which produce rapid heating and cooling cycles, but it can result in a service temperature limitation. Alpha prime is another embrittling phase that forms at slightly lower temperatures.

11.2 Constitution Diagrams

While carbon steels rely on the iron-iron carbide and CCT diagrams for predicting weld microstructures, stainless steels utilize what are known as "constitution" diagrams. A constitution diagram predicts the microstructure of the stainless steel based on the type and amount of various alloying elements. A variety of constitution diagrams have been developed over the years, but one of the very common ones still being used today is known as the Schaeffler diagram (Figure 11.2).

Constitution diagrams are based on formulas representing austenite stabilizing elements (nickel equivalency formula) on the vertical axis and ferrite stabilizing elements (chromium equivalency formula) on the horizontal axis. The reason these formulas are called nickel and chromium

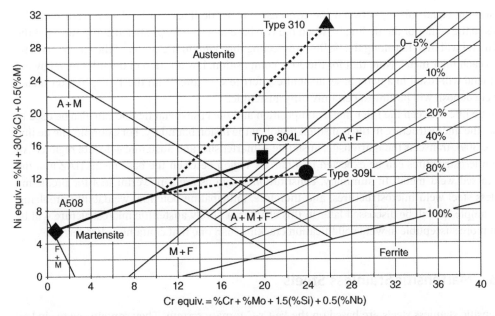

Figure 11.2 The Schaeffler diagram is ideal for predicting weld microstructures in a dissimilar metal weld between carbon and stainless steel. (*Source: Welding Metallurgy and Weldability of Stainless Steels*, Figure 9.1, page 290. Reproduced with permission from John Wiley & Sons).

equivalency formulas is because nickel is the predominant austenite stabilizer while chromium is the predominant ferrite stabilizer. The equivalency formulas include alloying elements and appropriate multiplying factors, which render them "equivalent" in potency to either nickel or chromium in terms of phase-stabilizing potency. The main differences among the various constitution diagrams available are their equivalency formulas and the range of stainless steel compositions they can be used for.

To predict weld microstructures, nickel and chromium equivalency calculations can be determined for both the base metal and filler metal and the two points plotted on the diagram. A tie line is then drawn between the two points; the predicted weld metal microstructure will lie somewhere along this line dictated by the amount of dilution of the weld filler metal by the base metal. For example, if the base metal dilution is 25%, the expected weld metal microstructure would fall along the tie line at a position located 25% of the entire length of the line closest to the weld metal composition. Dilution of the filler metal by the base metal is often an important consideration when welding certain stainless steels. For example, excessive base metal dilution in austenitic stainless steels is known to increase the susceptibility to solidification cracking, discussed later in this chapter.

Relative to all the available constitution diagrams, the Schaeffler diagram is useful in determining the "big picture" for stainless steel weld microstructures since a wide range of nickel and chromium equivalency is covered. It is also an ideal diagram for predicting weld microstructures in a dissimilar metal weld between stainless and carbon steels since both stainless steels and low alloy steels can be plotted on the diagram. Tie lines can be used for predicting dissimilar metal weld microstructures as well (again, refer to Figure 11.2). In the example shown, a low alloy steel such as A508 (diamond symbol) is being welded to Type 304L (square symbols) using either a Type 309L (round symbol) or Type 310 (triangular symbol) filler metal.

Assuming equal mixing between the two base metals, tie lines can be drawn from the filler metals to the center of the tie line between the two base metals. The predicted composition of the weld metal will then fall along the tie line connecting this point in the center of the base metal's tie line to each filler metal. Note that for the 309L composition, low dilution of the filler metal by the base metal will result in a two phase, austenite + ferrite structure. With the Type 310 filler metal, the weld deposit will be almost certainly fully austenitic.

The tie line between the filler metal composition and the base metal tie line can also be used to predict the microstructure in the transition region at the fusion boundary. For example, the tie line to Type 309L transects the martensite, austenite + martensite, and austenite + ferrite regions. All these microstructures can be expected in a narrow region between the base metal HAZ and the fully mixed weld metal. In summary, constitution diagrams are a powerful tool for predicting stainless steel weld microstructures, and therefore, anticipating potential weldability problems. For example, as will be discussed later in this chapter, a purely austenitic microstructure will be much more susceptible to weld solidification cracking than one that contains 5–10% ferrite.

11.3 Martensitic Stainless Steels

Martensitic stainless steels are based on the Fe-Cr-C ternary system. They contain relatively low amounts of chromium (12–18%) and high amounts of carbon (0.1–0.25% for most alloys, but up to 1.2% for cutlery grades). They undergo an allotropic transformation to austenite, and then form martensite upon cooling from the austenite. Martensite forms easily with these alloys, even at relatively slow cooling rates, so there is no need to make use of CCT diagrams when analyzing them. In addition to the dominate phase of martensite, they may also contain small amounts of ferrite and carbides.

A wide range of strengths are achievable with martensitic stainless steels. Yield strengths ranging from 40 ksi (275 MPa) in an annealed condition to 280 ksi (1900 MPa) in the quenched and tempered condition (for high carbon grades) are possible. When producing these steels, tempering is required to achieve acceptable toughness and ductility for most engineering applications. High hardness levels are also achievable, promoting abrasion resistance.

In general, corrosion resistance of martensitic stainless steels is not as good as the other grades due to the relatively low chromium content of most alloys. These alloys are generally selected for applications where a combination of very strength and reasonable corrosion resistance is required. The low chromium (for most grades) and alloying content of the martensitic stainless steels also makes them less costly than many other stainless steels.

Common applications of martensitic stainless steels include steam, gas, and jet engine turbine blades, and steam piping. The high chromium, high carbon grades are used to make items such as surgical instruments, knives, gears, and shafts. Martensitic stainless steels are not used above 650°C (1200°F) due to degradation in both mechanical properties and corrosion resistance. They will also begin to transform back to austenite at these temperatures. Because of the formation of untempered martensite during cooling after welding which increases the risk for hydrogen cracking, the martensitic stainless steels almost always require some combination of preheat, interpass control (maintaining a minimum temperature during multipass welding), and a postweld temper heat treatment. For this reason, they are generally considered to be the most difficult to weld among all the stainless steels.

The greatest risk for hydrogen cracking occurs when welding thick cross-sections and steels containing high carbon (>0.3 wt.%) contents. Therefore, the approach to preheat and interpass

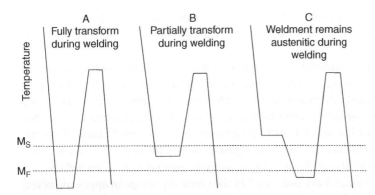

Figure 11.3 Different heat treatment approaches when welding martensitic stainless steels. (*Source: Welding Metallurgy and Weldability of Stainless Steels*, Figure 4.15, page 75. Reproduced with permission from John Wiley & Sons).

control can vary, depending on the thickness and the carbon content of the part being welded. Figure 11.3 shows three different preheat and interpass control approaches. In approach A, the weld region is allowed to cool below the martensite finish temperature resulting in a complete transformation to martensite. This will allow for the martensite to be tempered by a postweld tempering heat treatment, or by subsequent weld passes if it's a multipass weld. This method can be used for thinner cross-sections and lower carbon contents, when the risk of the martensite cracking before it is tempered is lower. Example B shows a preheat and interpass approach that results in only partial transformation to martensite. The risk of this approach is that some austenite will remain during the tempering treatment, and that austenite will then transform to untempered martensite upon cooling. Example C is often the recommended approach when the risk of hydrogen cracking during welding is the highest due to thick cross-sections and higher carbon contents. In this case, interpass temperature control (in a multipass weld) always maintains a temperature above the martensite start temperature until the final weld is made. The weldment is then allowed to cool slowly to transform entirely to martensite, and then given a final temper heat treatment.

Another issue when welding martensitic stainless steel can be the formation of small amounts of ferrite in both the fusion zone and HAZ. Since ferrite is very soft, it can reduce the strength of the weldment, even when it exists in small amounts. The presence of ferrite will be most likely when welding alloys with very low carbon contents since the size of the gamma loop will be reduced, so choosing a higher carbon content can eliminate the ferrite. A constitution diagram known as the Balmforth diagram can be used to predict the likelihood of forming ferrite when welding.

11.4 Ferritic Stainless Steels

Ferritic stainless steels are Fe-Cr alloys that contain sufficient chromium and small amounts of carbon (usually ≤0.12 wt.%) such that little or no austenite forms at elevated temperatures, which in turn, results in a microstructure that is primarily ferrite. These alloys are available in a wide range of chromium contents, and therefore, a wide range of corrosion resistance. Ferritic stainless steels are typically chosen when corrosion resistance is more important than high strengths. The low chromium containing versions (11–12 wt.% or less) tend to be among the least expensive of all the stainless steels.

Since ferritic stainless steels are essentially a single-phase alloy up to the melting point, they are not transformation hardenable. They have a body-centered cubic crystal structure so they have poor toughness at low temperatures and exhibit a ductile-to-brittle transition temperature (DBTT), much like carbon steels. This transition behavior makes them unsuitable for low temperature service, while service temperatures can be as high as 400°C (750°F). At temperatures above 400°C, these alloys may become brittle due to the formation of alpha-prime or sigma phase (see the Fe-Cr diagram in Figure 11.1). But because these phases form very slowly, they are not typically a problem when welding.

Historically, ferritic stainless steels have been used in the greatest tonnage in applications that do not require welding. For example, the medium chromium grades are used extensively for automotive trim and other decorative/architectural applications. Over the past 20 years or so, the use of low and medium chromium grades for welded automotive exhaust systems has increased dramatically. This has effectively eliminated the need to routinely replace automotive exhaust systems that historically were made of carbon steel and corroded quite rapidly. A common modern example is the use of low chromium 409 stainless steel, which is routinely welded using High Frequency Resistance Welding to create exhaust tubing (Figure 11.4).

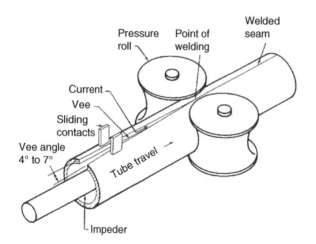

Figure 11.4 High frequency resistance welding is routinely used to weld 409 stainless steel to produce automotive exhaust tubing. (*Source:* Reproduced by permission of American Welding Society, ©*Welding Handbook*).

Many high-Cr grades have been developed for use in demanding environments, such as chemical plants, pulp and paper mills, and refineries. These alloys possess superior corrosion resistance relative to the austenitic and martensitic grades. However, they are relatively expensive and difficult to fabricate. The weldability of alloys with greater than 25% chromium has been the subject of considerable research. Because ferritic stainless steels are single-phase at elevated temperatures, grain growth (Figure 11.5) during welding can be quite rapid, resulting in degradation to both ductility and toughness. As a result, ferritic stainless steels are known to suffer from losses in ductility and toughness during welding. Increasing amounts of carbon, chromium, nitrogen, and impurity elements accentuate the problem. Approaches to welding include keeping heat input low during welding and preheat temperatures (if used) to a minimum. Titanium and niobium may be added to control the carbon and nitrogen. Multipass weldments may also experience excessive grain growth, resulting in the degradation of mechanical properties in the HAZ and weld metal.

Figure 11.5 Ferritic stainless steel (409) weld microstructure revealing extremely large HAZ grain sizes (left) near the fusion zone.

Figure 11.6 Alloy 409 microstructure consisting of mostly ferrite with small amounts of martensite.

Although these alloys are mostly ferritic, small amounts of grain boundary martensite may form (Figure 11.6) in some alloys (such as 430) since the gamma loop (austenite) cannot be completely avoided during cooling. These alloys may therefore be susceptible to hydrogen cracking, mandating the use of low hydrogen practice and possibly the need for a postweld heat treatment.

11.5 Austenitic Stainless Steels

Austenitic stainless steels represent the largest group of stainless steels and are produced in greater quantities than any other grade. The microstructures of these steels are mostly austenite but may contain small amounts of ferrite. They exhibit excellent corrosion resistance in most environments, with strengths equivalent to mild steels. Austenitic stainless steels also offer a good balance of toughness and ductility but are more expensive than the martensitic and most of the ferritic grades due to the higher alloy content of these alloys. Because they contain elements such as nickel that stabilize austenite down to room temperature, they are not transformation hardenable. Due to a face-centered cubic crystal structure, their low temperature impact properties are quite good, making them applicable for cryogenic applications. Service temperatures can be as high as 760°C (1400°F). Despite their higher cost, they offer many distinct engineering advantages over other stainless steels, including corrosion resistance, high temperature performance, formability, and weldability (when proper procedures are followed).

Elements that stabilize austenite, most notably nickel, are added in large quantities to these steels. They contain between 8 and 20% nickel, and between 16 and 25% chromium. Carbon contents are relatively low, ranging between 0.02 and 0.08%. In addition to nickel, nitrogen and copper are strong austenite stabilizing elements, as evident from the various nickel equivalency formulas. Nitrogen also may be added to significantly improve strength. Nitrogen-strengthened alloys are usually designated with a suffix N added to their AISI designation (such as 304LN), or under various trade names such as Nitronic®.

Austenitic stainless steels are used in a wide range of applications (Figure 11.7) including structural support and containment, pressure vessels, architectural, kitchen equipment, and medical products. Some of the more highly alloyed grades are used in very high temperature service for applications such as heat-treating baskets. While austenitic stainless steels are generally very resistant to corrosion, they are not good choices in extreme cases such as highly caustic or chloride-containing environments such as seawater. This is due to their susceptibility to stress corrosion cracking, a phenomenon that afflicts the base metal, HAZ, and weld metal.

Figure 11.7 Austenitic stainless steels are used for a wide range of applications, from architectural applications such as the St. Louis Arch. (*Source:* Daniel Schwen Derivative Work) to kitchen sinks (*Source:* Kohler).

Although the austenitic alloys are generally considered to be very weldable, they are subject to many weldability problems if proper precautions are not taken. Weld solidification cracking is one of the biggest concerns, particularly with those alloys that solidify as 100% austenite. Despite the good general corrosion resistance of these alloys, they may also be subject to localized forms of corrosion at grain boundaries in the HAZ or at stress concentrations in and around the weld.

Other weldability problems are possible but less common, including reheat cracking, ductility dip cracking, HAZ liquation cracking, and copper contamination cracking. Because a common approach for eliminating solidification cracking is to use filler metals that promote weld metal ferrite formation, intermediate temperature embrittlement due to sigma phase formation is also a concern. As with the ferritic stainless steels, the sigma phase precipitation reaction is relatively sluggish, and is usually a service-related rather than a welding-related problem. Although not considered a weldability problem, austenitic stainless steels are known to produce significant distortion when welded due to the very high coefficient of thermal expansion of these alloys.

As mentioned, a primary weldability problem with austenitic stainless steels is solidification cracking, which occurs when liquid-containing solidification grain boundaries pull apart as the weld cools and shrinks during the final stages of weld solidification. The likelihood for solidification cracking is a strong function of composition, as shown by the plot (Figure 11.8) of cracking susceptibility versus Cr_{eq}/Ni_{eq} ratio. The equivalency formulas used to determine this ratio are the same formulas used for the constitution diagrams and establish the relative amounts of ferrite (Cr_{eq}) and austenite (Ni_{eq}) stabilizing elements. The letters on the plot represent variations in the weld solidification mode ranging from fully austenitic (A) to fully ferritic (F). Note that compositions that result in primary austenite solidification (A and AF) are most susceptible to solidification cracking (Figure 11.9), while the FA mode offers the greatest resistance. This is because the presence of a two phase (ferrite + austenite) microstructure at the end of solidification resists wetting along solidification grain boundaries, effectively reducing the extent of the liquated grain boundaries that are prone to cracking.

Filler metal selection can be a very effective means for controlling weld solidification cracking with austenitic stainless steels. A typical approach is to use filler metals that contain enough ferrite-stabilizing elements to promote the FA solidification mode. On a typical constitution

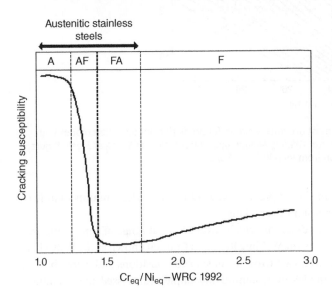

Figure 11.8 Solidification cracking susceptibility plot reveals the advantage of solidifying as ferrite. (*Source: Welding Metallurgy and Weldability of Stainless Steels*, Figure 6.21, page 175. Reproduced with permission from John Wiley & Sons).

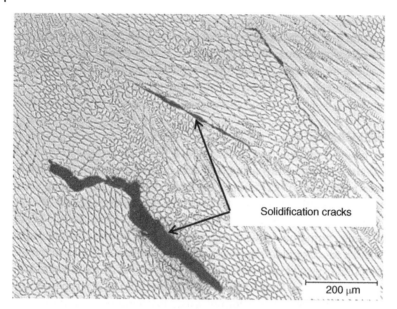

Figure 11.9 Solidification as 100% austenite ("A" mode) significantly increases the susceptibility to weld metal solidification cracking. (*Source:* Lippold et al. (2005), Reproduced with permission from John Wiley & Sons).

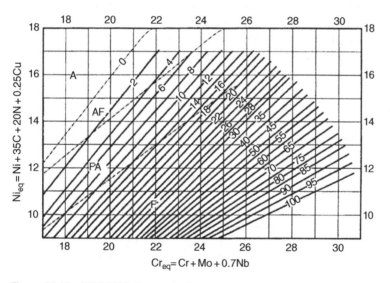

$$Cr_{eq}= Cr + Mo + 0.7Nb$$

Figure 11.10 WRC-1992 diagram is the most accurate diagram for predicting solidification modes with austenitic and duplex stainless steels. (*Source: Welding Metallurgy and Weldability of Stainless Steels*, Figure 3.14, page 42. Reproduced with permission from John Wiley & Sons).

diagram such as the Schaeffler diagram shown previously, weld metal ferrite amounts of at least 5–10% are generally considered to be resistant to cracking.

Since the introduction of the Schaeffler diagram, which covers a broad range of compositions, a more recent diagram has been developed by the Welding Research Council known as the WRC-1992 diagram (Figure 11.10). This diagram covers a narrower range of compositions and is much more accurate for predicting solidification modes in austenitic and duplex (discussed next) stainless steels. It is now widely accepted worldwide as a valuable source for predicting susceptibility to solidification cracking. Other factors increasing the susceptibility to solidification cracking include high amounts of impurities such as phosphorus, boron, and sulfur, fast travel speeds that promote more of a teardrop (vs. elliptical) shaped weld, and high weld restraint.

Another common weldability problem with austenitic stainless steels known to result in a severe loss in corrosion resistance in the HAZ is called sensitization. In a narrow portion of the HAZ heated to temperatures ranging from 600 to 850°C (1110 to 1560°F), chromium and carbon combine to form carbides ($Cr_{23}C_6$) along grain boundaries. At temperatures above this range, carbides formed during heating are dissolved, and temperatures below this range are not sufficient for carbide precipitation. This results in a very narrow region of carbide formation.

As the chromium-rich carbides form, they extract chromium from the region immediately adjacent to the grain boundaries (Figure 11.11). When the chromium content in these regions drops below 12%, it is effectively no longer a stainless steel. If exposed to a corrosive environment, grain boundary corrosion will occur along this very narrow region in the HAZ (Figure 11.12). This region is also susceptible to a form of cracking known as stress corrosion cracking.

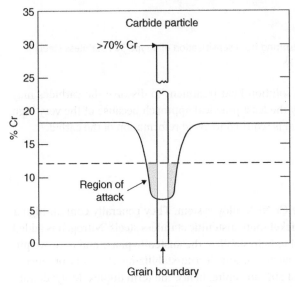

Figure 11.11 Chromium depletion along HAZ grain boundaries can result in sensitization. (*Source:* Dr. Mary Juhas, The Ohio State University).

Approaches to eliminating sensitization include using low carbon or stabilized grades of stainless steel. The low carbon grades are designated with the letter L, for example, 304L and 316L, and are less susceptible to sensitization due to the reduced carbon, which slows the rate by which chromium carbides can form relative to conventional grades. Stabilized grades include 347 and 321, and are similar in composition to 304 except for the additions of titanium and niobium. These elements form more stable carbides than the chromium carbide, and effectively "tie-up" the carbon preventing chromium carbides from forming. In summary, both the low carbon and stabilized grades offer a similar benefit, which is a reduction in chromium carbide formation. By reducing the amount of carbide formation, the chromium depletion from the matrix and risk of sensitization is reduced.

Other options for avoiding sensitization include utilizing low weld heat inputs to minimize the time in the carbide participation sensitization range. Approaches to rapidly cooling the weld may

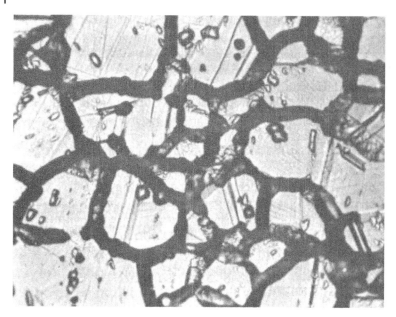

Figure 11.12 HAZ grain boundary corrosion resulting from sensitization of a type 304 stainless steel. (*Source*: Dr. Mary Juhas, The Ohio State University).

be considered as well. Finally, a postweld solution heat treatment to dissolve the carbides may solve the problem. However, this is usually the least practical approach because of the very high solution temperatures and rapid cooling that is required to avoid reformation of the carbides.

11.6 Duplex Stainless Steels

Duplex stainless steels are based on the Fe-Cr-Ni-N alloy system. They generally contain higher amounts of chromium (22–32%) and less nickel than austenitic stainless steels. Nitrogen is added (up to 0.25%) as an important alloy element to help stabilize the austenite phase and improve pitting corrosion resistance. The balance of austenite and ferrite stabilizing elements produce a microstructure of nominally 50% ferrite and 50% austenite, hence the term duplex. Molybdenum and copper are also added to some alloys to improve corrosion resistance. They are usually designated using a four number system with the first two numbers indicating the amount of chromium added and the second two indicating the amount of nickel. For example, alloy 2507 contains 25 wt.% chromium and 7 wt.% nickel. Because of the higher alloy contents and meticulous processing required, the duplex stainless steels are more costly to produce than the austenitic alloys but offer distinct corrosion advantages and weight savings. The duplex alloys can be nearly double the strength of austenitic stainless steels. Their development has evolved rapidly over the past 20–25 years, and they are being introduced in greater quantities and wider applications. Significant improvements have been made in both the weldability and corrosion resistance of these alloys over this period.

Duplex stainless steels are used in applications that take advantage of their superior corrosion resistance, strength, or both. They also have a lower thermal expansion coefficient than austenitic stainless steels, which can be an advantage in many applications. Their superior corrosion resistance relative to austenitic stainless steels is especially important in very caustic or seawater containing environments, and in applications with exposure to hydrogen sulfide. They perform particularly well when stress corrosion cracking and pitting corrosion are concerns, and are also vastly superior to structural steels in most corrosive applications while offering comparable strength. Because duplex stainless steels form many embrittling precipitates at relatively low temperatures, they are not recommended for service applications exceeding about 300°C (570°F). They may be used in place of some Ni-based alloys, at a fraction of the material cost. Applications include pipelines, offshore oil production umbilical systems, chemical plants, and pulp and paper mills.

Probably the biggest challenge associated with welding these alloys is maintaining the ferrite-to-austenite phase balance, which plays a primary role in their attractive properties. Altering this balance will affect their corrosion resistance, ductility, and toughness, so it is essential to control welding heat inputs and cooling rates. A common problem that can result from rapid cooling rates during welding is to create a microstructure that contains excessive amounts of ferrite. This is because there may not be enough time for austenite formation in the solid-state during cooling. High amounts of ferrite can reduce toughness and corrosion resistance, and result in a microstructure that is susceptible to hydrogen cracking, possibly mandating the use of low hydrogen practice. Figure 11.13 reveals a duplex stainless steel fusion zone with an acceptable balance of ferrite (light shaded) to austenite (dark shaded).

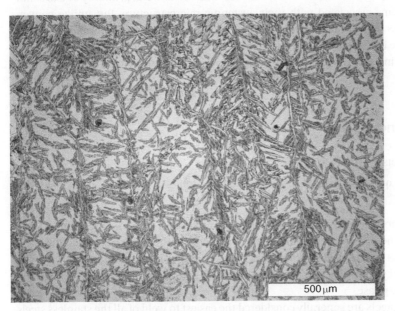

Figure 11.13 Duplex stainless steel fusion zone microstructure—ferrite is light shaded, austenite is dark.

Welding with higher heat input processes to achieve slower cooling rates is one approach to maintaining the ferrite-to-austenite balance. Another approach involves the use of "nickel-boosted" filler metals which stabilize the austenite, but of course this only affects the fusion zone. To ensure that the HAZ retains a proper balance of ferrite and austenite, the best approach may be to choose a base metal alloy composition that is less likely to develop a phase imbalance during welding. Many of the modern duplex alloys are designed just for this purpose. Because they solidify as bcc ferrite, duplex stainless steels are generally not susceptible to solidification cracking unless impurity levels are high.

11.7 Precipitation-Hardening Stainless Steels

The precipitation-hardening (PH) stainless steels represent a smaller group of stainless steels that derive additional strength from elements added such as copper and niobium which provide for the ability to precipitation strengthen. These steels are grouped into three categories based on their primary microstructure—martensitic, austenitic, and semiaustenitic. Corrosion resistances are similar to austenitic stainless steels, while strengths can exceed 1380 MPa (200 ksi). Applications include valves, gears, pressure vessels, jet engine blades and components, and rocket engines. They are designated by either a two-number system based on their chromium and nickel content (i.e., 17–4 PH contains 17 wt.% chromium and 4 wt.% nickel) or a trade name. Due to their complex processing requirements they tend to be very expensive, and therefore, are typically only used for critical and demanding applications.

The PH stainless steels are typically welded in an annealed or solutionized condition, and then given a postweld heat treatment. In the case of the martensitic alloys, the postweld heat treatment can be a single heat treatment that accomplishes both precipitation hardening and tempering of the martensite. The postweld heat treatment for the semiaustenitic alloys is more complex and may involve multiple steps to recover the base metal mechanical properties. Of the three groups, the austenitic alloys are usually considered to be the most difficult to weld, primarily due to solidification cracking problems associated with their austenitic microstructure. But unlike conventional austenitic stainless steels, welding with filler materials to promote ferrite formation is not a feasible solution because the weld metal cannot be precipitation strengthened. Therefore, the best approaches when welding these alloys are to keep impurity elements and weld stresses to a minimum.

11.8 Test Your Knowledge

I) Fundamental Concepts—True/False

The following true/false questions pertain to some of the most important fundamental concepts in this chapter:

1) Stainless steels are known to contain a minimum of approximately 12 wt.% chromium.
2) Chromium is the element known to expand the gamma loop on the Fe-Cr phase diagram.
3) Constitution diagrams can be used to predict the microstructures of stainless steels.
4) The nickel equivalency formulas used on constitution diagrams include elements known to stabilize the austenite phase.
5) Martensitic stainless steels are generally considered the easiest to weld of all the stainless steels.
6) Ferritic stainless steels are known to form extremely small grain sizes in the weld HAZ, and therefore, have excellent toughness.
7) A common approach to mitigating solidification cracking when welding austenitic stainless steels is to use filler metals that promote the formation of some ferrite.
8) A corrosion problem known as sensitization that can occur when welding some austenitic stainless steels is due to localized depletion of chromium along the HAZ grain boundaries.
9) A primary problem when welding duplex stainless steels is the formation of a microstructure containing excessive amounts of ferrite.

II) Solve a Welding Engineering Problem

The following challenge represents a typical problem Welding Engineers might encounter in their career:

Your latest challenge is to develop an approach to welding a certain martensitic stainless steel. A glance at the chemical composition of this alloy tells you that the carbon content is 0.60 wt.%, and you also know that the fabrication your company will produce with this stainless steel has very thick cross-section. Please describe your welding concerns and your recommended approach to welding this fabrication.

Recommended Reading for Further Information

ASM Handbook, Tenth Edition, Volume 6—"Welding, Brazing, and Soldering," ASM International, 1993.

AWS Welding Handbook, Ninth Edition, Volume 4—"Materials and Applications, Part 1," American Welding Society, 2011.

Welding Metallurgy and Weldability of Stainless Steels, John Wiley & Sons, Inc., 2005.

12

Welding Metallurgy of Nonferrous Alloys

12.1   Aluminum Alloys

Aluminum alloys represent a family of widely used engineering materials in applications that usually require a combination of low density (light weight) and corrosion resistance. The corrosion resistance of these alloys results from the rapid formation of an aluminum oxide (Al_2O_3) on the surface which is relatively stable at ambient temperatures. Because of this beneficial characteristic, applications involving aluminum often do not require protective paint or coatings. Unlike steels, they do not exhibit a ductile-to-brittle transition at low temperatures. The use of aluminum has increased markedly since the early 1990s. Most alloys are easily rolled, extruded, and drawn, and can be fabricated into a variety of shapes. With the recent thrust toward more lightweight electric and fuel-efficient and vehicles, the increase in potential aluminum applications in the automotive industry means the need to weld these alloys will continue to grow.

Eight classes of wrought aluminum alloys have been defined by the Aluminum Association (Table 12.1). Individual alloys are designated using a four-digit number, with the first digit representing the basic class. The classes are distinguished by the primary alloying element(s) and how they are strengthened—either by cold work (strain hardening) or a precipitation-strengthening heat treatment. Other digits in the designation have no special significance. As the table shows, the 1xxx, 3xxx, 4xxx, and 5xxx alloys are all strengthened by cold work, whereas the 2xxx, 6xxx, 7xxx, and 8xxx alloys are all strengthened by a precipitation heat treatment. It is important to recognize the strengthening method for the alloy prior to welding. For both strengthening methods, considerable degradation of the tensile strength in the weld HAZ can be expected, but it is possible to recover strength when welding alloys that are strengthened by precipitation-strengthening.

The four-digit alloy code is often followed by a heat treatment, processing, or temper designation code. This code provides some detail regarding how the alloy is strengthened. For example, the "T6" in 6061-T6 refers to a solution heat treatment followed by a precipitation-strengthening aging treatment, while the "H3" in 5754-H3 refers to a strain hardening followed by a low-temperature stabilization treatment. These are just two examples of many possible temper designation codes. The word "temper" is very loosely used with aluminum alloys since in some cases there is no heat treatment involved at all. This version of the word "temper" should not be confused with the heat treatment associated with softening martensite in steels.

Aluminum alloys are used in a wide number of applications ranging from decorative to structural. Because of their relatively high strength, the 2XXX and 7XXX series alloys are used extensively in aerospace applications. Most of the wing and fuselage "skin" on commercial airliners use

Welding Engineering: An Introduction, Second Edition. David H. Phillips.
© 2023 John Wiley & Sons, Inc. Published 2023 by John Wiley & Sons, Inc.
Companion Website: www.wiley.com/go/Phillips/WeldingEngineeringIntroduction

Table 12.1 The eight families of aluminum alloys.

Series	Type of alloy composition	Strengthening method	Tensile strength range	
			MPa	ksi
1xxx	Al	Cold work	70–175	10–25
2xxx	Al-Cu-Mg (1–2.5% Cu)	Heat treat	170–310	25–45
2xxx	Al-Cu-Mg-Si (3–6% Cu)	Heat treat	380–520	55–75
3xxx	Al-Mn-Mg	Cold work	140–280	20–40
4xxx	Al-Si	Cold work (some HT)	105–350	15–50
5xxx	Al-Mg (1–2.5% Mg)	Cold work	140–280	20–40
5xxx	Al-Mg-Mn (3–6% Mg)	Cold work	280–380	40–55
6xxx	Al-Mg-Si	Heat treat	150–380	22–55
7xxx	Al-Zn-Mg	Heat treat	380–520	55–75
7xxx	Al-Zn-Mg-Cu	Heat treat	520–620	75–90
8xxx	Al-Li-Cu-Mg	Heat treat	280–560	40–80

7XXX series alloys. The 5XXX and 6XXX series alloys are considered the "workhorse" alloys and are used extensively for structural applications. Most of the aluminum currently used for automotive applications is in either the 5XXX or the 6XXX alloy family. The 8XXX series alloys contain additions of lithium to reduce density and improve strength resulting in excellent strength-to-weight ratios that provide obvious advantages for aerospace applications. The 1XXX and 3XXX series alloys represent the largest tonnage production of aluminum alloys since they are used for aluminum foil and beverage cans, respectively. Due to the high electrical conductivity of aluminum, the 1XXX alloys are also used in a variety of electrical applications, such as for high-voltage transmission lines.

As mentioned previously, when welding all aluminum alloys, significant degradation in hardness and tensile strength can be expected in the HAZ (Figure 12.1). In many cases, as much as 50% reduction in properties will occur. Softening occurs in the HAZ of cold worked aluminum alloys due to recrystallization and grain growth. There is no way to recover this loss in properties unless the material is cold worked again, which is rarely a practical solution. As a result, cold worked aluminum alloys are often considered unweldable, not because they cannot be welded but because of the drastic loss in unrecoverable properties. The best approach is to keep heat input very low to

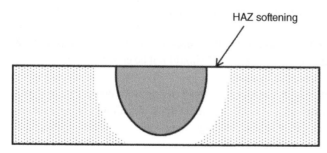

Figure 12.1 All aluminum alloys suffer from significant degradation to hardness and tensile strength in the weld HAZ.

minimize the loss in strength, or to weld in the fully annealed condition and compensate for the lower strength annealed condition in the design of the part or fabrication.

The weld HAZ softening that occurs when welding precipitation-strengthened alloys is due to overaging and/or dissolution of the strengthening precipitates that were formed during the precipitation hardening treatment. When welding these alloys, however, there is an opportunity to recover some or all the properties with a postweld heat treatment. One approach is to solutionize the entire weldment to completely dissolve the precipitates, followed by an aging treatment. This approach is effective but time consuming and costly, and in many cases, not practical. Another approach is to weld in the "T4" condition. This condition includes a "natural" aging treatment that does not fully strengthen the alloy. When an alloy is welded in the "T4" condition, the extent of overaging is reduced. The weld can then be aged to achieve near base metal strength levels.

Aluminum is well known for its propensity to form weld metal porosity (Figure 12.2). This is because molten aluminum can dissolve considerable amounts of hydrogen, especially at higher temperatures. As indicated in Figure 12.3, once aluminum melts (660°C/1220°F), its solubility for hydrogen increases dramatically. The solubility then increases rapidly upon further heating of the

Figure 12.2 Porosity in the weld metal of an aluminum weldment.

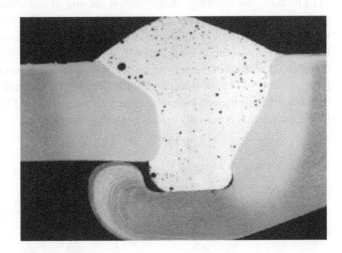

Figure 12.3 The rapid increase in solubility of hydrogen in molten aluminum results in the propensity for forming weld metal porosity.

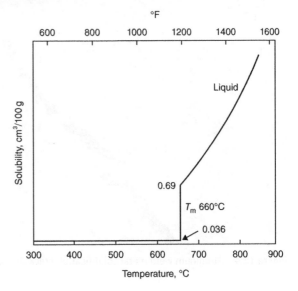

liquid aluminum. During welding, as the weld metal cools, hydrogen solubility drops rapidly causing it to come out of solution and form gas bubbles. Upon further cooling and subsequent solidification, the bubbles become trapped resulting in porosity.

When proper procedures are followed, the porosity is generally small and well dispersed, and may be acceptable per the applicable code. Total elimination of weld metal porosity in aluminum weldments is extremely difficult, and in most cases, virtually impossible. And because of extremely high solubility for hydrogen in lithium, the aluminum lithium alloys may be even more susceptible to porosity.

Reduction of weld metal porosity is accomplished by removing all possible sources of hydrogen. A major source of hydrogen is moisture contained in the surface oxide layer, which can easily be removed by mechanical methods such as machining, scraping, or wire brushing. Welding should commence within 24 hours of removing the oxide since it forms so rapidly. Other sources of hydrogen that should be removed include cleaning fluids, oils, or grease. Electromagnetic stirring of the weld puddle has been studied as an approach to promote hydrogen bubbles rising and exiting the surface of the weld pool. However, this approach is not very practical.

Aluminum alloys are also very susceptible to weld solidification cracking, (described in Chapter 11) a form of "hot cracking" that occurs at the end of solidification due to the presence of liquid grain boundary films. This susceptibility is driven primarily by the wide solidification temperature range of most aluminum alloys, and the subsequent contraction stresses that develop due to the large coefficient of thermal expansion (CTE) of aluminum. Solidification cracking can be minimized or eliminated in many cases by controlling the weld metal microstructure. The presence of large amounts (>10%) of eutectic liquid at the end of solidification can significantly reduce cracking through a mechanism known as eutectic healing. Figure 12.4 shows an aluminum weld solidification crack that is partially healed by this mechanism.

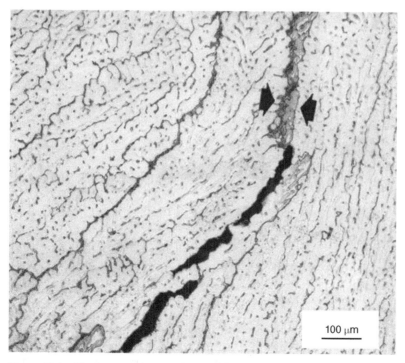

100 μm

Figure 12.4 Aluminum weld metal solidification crack that is partially healed by eutectic liquid.

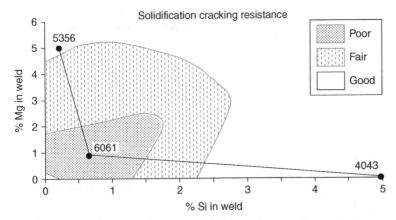

Figure 12.5 Aluminum weld filler metals such as 5356 and 4043 greatly increase solidification cracking resistance by promoting the formation of eutectic liquid.

The tendency for solidification cracking increases with the presence of alloying elements such as copper, magnesium, and zinc. However, as the levels of these elements are increased, cracking may be mitigated by eutectic healing. Therefore, solidification cracking is often remedied by welding with a filler metal that has greater alloying content than the base metal (Figure 12.5) to promote eutectic healing. As the figure shows, alloy 6061, which is very susceptible to solidification cracking, can be effectively welded using filler metals containing high amounts of either magnesium (5356) or silicon (4043). However, since these filler metals are not precipitation hardenable, their use will limit the strength of weldments made with alloys such as 6061. Another method for reducing solidification cracking susceptibility is minimizing restraint, but utilizing an appropriate filler metal is usually the best approach.

12.2 Nickel-Based Alloys

Nickel-based alloys exhibit many desirable properties that make them ideal candidates for high-performance applications. They are most often selected for their outstanding corrosion resistance, and/or their combination of strength and corrosion resistance, particularly at elevated temperatures. Strengthening of their austenitic (FCC) microstructure is achieved by either solid solution or precipitation strengthening. Machining of these alloys is relatively difficult as compared to steels. Because of their high alloy content and complex processing methods, they are very expensive (10–20 times more expensive than carbon steel and 3–4 times more expensive than stainless steel) and are normally selected only for specialty applications where other metals would not survive.

Nickel-based alloys are generally defined as those alloys in which Ni is the primary element. However, in some cases, the total of all alloying additions may exceed 50%, but if the Ni content represents the highest percentage of a single element, the alloy is still considered to be nickel-based. Some alloys are included among the family of nickel-based alloys even though they contain large amounts of iron. For example, Incoloy alloys 800 and 825 are actually iron-based alloys. There is a wide range of nickel-based alloys to choose from, and many are quite complex regarding the amount and number of alloying additions. Figure 12.6 shows some common alloy families as a function of Ni content. Many nickel-based alloys are known by their trade names such as Inconel, Nimonic, Waspaloy, René, Sanicro, and so on.

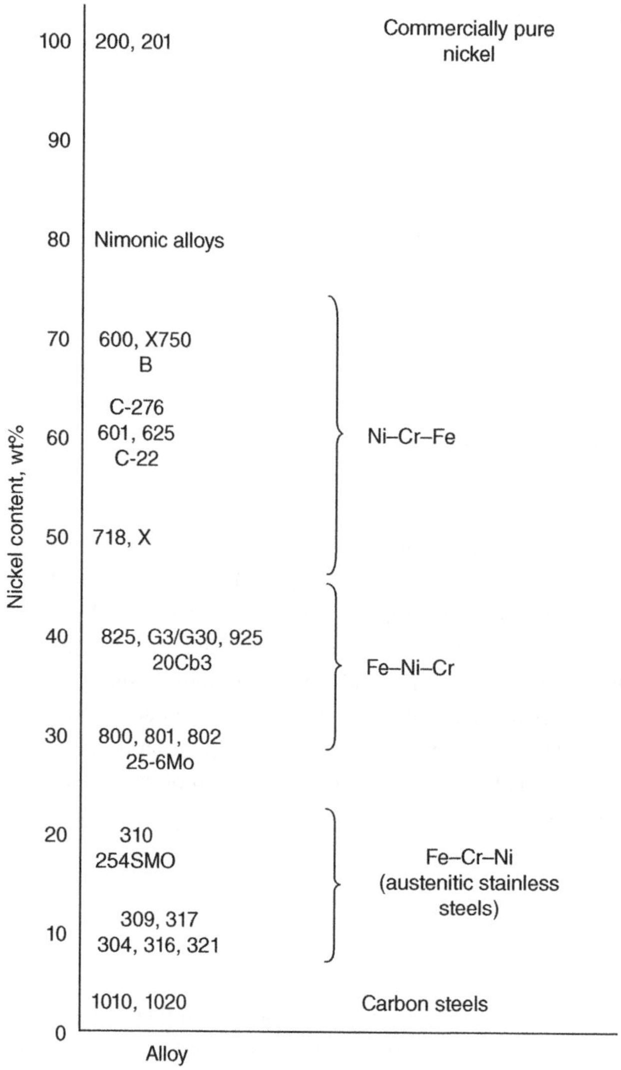

Figure 12.6 Classes of alloys as a function of nickel content.

Nickel-based alloys are generally grouped into one of two categories based on the method used to strengthen them—either solid solution strengthened, or precipitation hardened. Solid solution-strengthened alloys contain large additions of Cr, Mo, Fe, and occasionally W. The Cr and Mo also provide additional corrosion resistance. Solid solution-strengthened alloys 600 and 625 are widely used in the power generation and chemical processing industries. The precipitation hardened alloys are often called "superalloys" because of the exceptional strength they possess and maintain to very high temperatures. For this reason, they are widely used in gas turbine engine applications (Figure 12.7) at temperatures exceeding 650°C (1200°F). The superalloys can be strengthened by a variety of precipitates, but the most common is known as gamma prime, $Ni_3(Ti,Al)$.

In addition to gas turbine engines, applications for nickel-based alloys include nuclear pressure vessels and piping, heat exchangers, chemical processing, petrochemical, marine, and pulp-and-paper. They are often used for cladding, particularly to protect carbon steels in aggressive corrosion environments. Cladding operations can be cost-effective because they can produce fabrications with excellent corrosion resistance, but at a reduced cost compared to a fabrication made from 100% nickel-based material. Nickel-based alloys also exhibit a CTE that is midway between those of carbon steels and stainless steels. Therefore, they are sometimes used as a CTE mismatch "buffer" to reduce stresses during elevated temperature exposure of dissimilar metal carbon-to-stainless steel weldments. Because of their high cost, they generally see limited use in the mass production industries such as the automotive sector.

Nickel-based alloys are susceptible to a variety of weldability problems, but the most common are strain-age cracking, solidification cracking, and HAZ liquation cracking. Ductility-dip cracking has also been observed in some of the high chromium solid solution-strengthened alloys. Porosity is an occasional problem but can normally be controlled by proper cleaning procedures prior to welding.

Strain-age cracking is a particular problem with the precipitation-strengthened alloys (superalloys) that are strengthened to high levels by the rapidly forming gamma-prime precipitates. This type of cracking normally occurs during a postweld heat treatment and is due to

Figure 12.7 The gas turbine engine is a common application for nickel-based superalloys. (*Source:* GE Digital Energy).

the simultaneous formation of the strengthening precipitates and relaxation of residual stresses. For the superalloys, a postweld heat treatment is needed because during welding, the strengthening precipitates are dissolved in the HAZ. To recover the strength and reduce residual stresses, a postweld solution heat treatment followed by aging is required.

If the solutionizing heating rates during the postweld heat treatment are too slow, precipitation can occur prematurely (Figure 12.8). As the precipitates form within the grains, they strengthen the interior of the grain relative to the grain boundary while relaxation of residual stress occurs at the same time. Since the grain boundaries are weaker than the grain interiors, this can result in concentrated strains at or near the grain boundaries. If these strains are sufficiently high, grain boundary failure will occur and a strain-age crack will form. Thus, strain-age cracking takes its name from the simultaneous presence of both relaxation strains and a strong precipitation (aging) reaction.

Because strain-age cracking occurs due to the rapid formation of the $Ni_3(Ti,Al)$ precipitate upon reheating after welding, titanium and aluminum have an important influence on cracking susceptibility. By reducing the amounts of these elements, the rate of gamma-prime precipitation is reduced, effectively pushing the nose of the precipitation curve to longer times. Figure 12.9 shows the relative strain-age cracking effect of aluminum and titanium for various nickel-based alloys. Alloys with high titanium + aluminum contents, such as IN100 and IN713C, are almost impossible to postweld heat treat without causing strain-age cracking. Alloys such as Waspaloy that have intermediate titanium + aluminum contents have variable susceptibility to cracking, depending

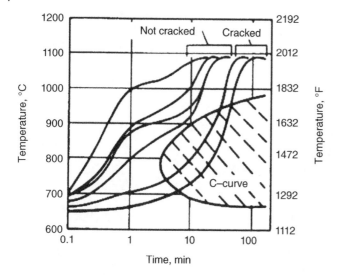

Figure 12.8 During a postweld heat treatment, nickel-based superalloys are susceptible to strain-age cracking, especially if heating rates are too slow. (*Source:* Berry and Hughes, *Welding Journal*, Vol. 46).

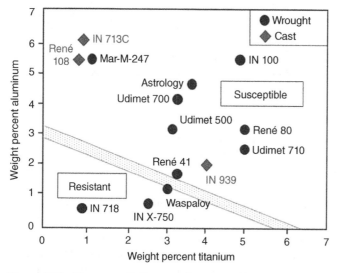

Figure 12.9 The susceptibility to strain-age cracking of nickel-based alloys increases with increases in aluminum and titanium. (*Source:* Adapted from Prager and Shira, WRC Bulletin 128).

on welding and restraint conditions. Alloy 718 (IN718) is mostly immune from strain-age cracking since it is alloyed with niobium to form a gamma double prime, Ni_3Nb precipitate. The strain-age cracking risk with this alloy is minimized because gamma double prime forms much more sluggishly than gamma prime, which allows stress relaxation to occur in the absence of precipitation.

Strain-age cracking is most often avoided by proper alloy selection. Figure 12.9 provides a guideline for choosing less susceptible alloys based on their aluminum and titanium content. However, as the Ti + Al additions are reduced, the volume fraction of gamma prime decreases,

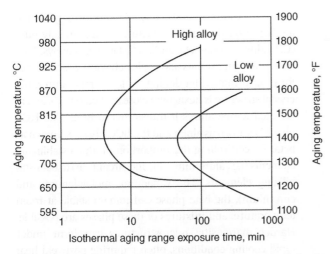

Figure 12.10 Nickel-based alloys with reduced amounts of aluminum + titanium result in the nose of the precipitation reaction curve moving to much longer times. (*Source:* Berry and Hughes, *Welding Journal*, Vol. 46).

and therefore, the high temperature strength of the alloy will be affected. This results in a trade-off between high temperature mechanical properties and weldability, but a balance of strength and reduced cracking sensitivity may be achieved by selecting an alloy with relatively low amounts of titanium and aluminum content. The precipitation reaction (Figure 12.10) upon reheating alloys with less Ti + Al in the temperature range where stress relaxation occurs will be retarded, and cracking may be avoided.

If selection of a less susceptible alloy is not possible, there may be an opportunity to avoid cracking through adjustments to the heat treatment. For example, per Figure 12.8, if heating rates to the solution temperature are rapid enough, the nose of the gamma double prime precipitation reaction can be avoided. However, this approach may not be practical with large fabrications. Another approach is to use a postweld heat treatment that imposes a hold time below the nose of the precipitation curve prior to heating to the solution temperature. This will provide some reduction in residual stresses to mitigate the stress relaxation that coincides with the precipitation reaction.

Since nickel-based alloys have fully austenitic microstructures, weld solidification cracking is also a potential problem, especially under high restraint conditions. HAZ liquation cracking can also occur, particularly in some of the superalloys. In most cases, solidification and liquation cracking susceptibility can be minimized by reducing both impurity concentrations and restraint.

12.3 Titanium Alloys

Titanium alloys exhibit low density and can be strengthened to high levels by a combination of solid solution and transformation hardening. As a result, they are known for their superior strength-to-weight ratio, or specific strength. In addition, they offer excellent corrosion resistance in most environments, including seawater. Some of the alloys can be used at temperatures as high as 540°C (1000°F). Titanium alloys are very expensive, primarily due to the very time-consuming and expensive chemical extraction process used to separate titanium from its ore (rutile).

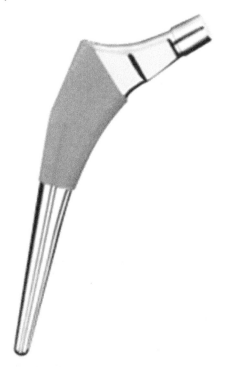

Figure 12.11 Medical devices and implants are often made from titanium. (*Source:* Zimmer Biomet).

There are four general classes of titanium alloys which are based on their microstructure—commercially pure (CP), alpha, alpha-beta, and metastable beta alloys. Like iron, titanium exhibits allotropic behavior, which allows for transformation hardening. At low temperatures, the crystal structure is hexagonal close packed (HCP), and at elevated temperatures it transforms to BCC. The HCP phase is known as alpha and the BCC phase is known as beta. By controlling the proportion of alpha-stabilizing elements (i.e., aluminum, tin, and zirconium) to beta-stabilizing elements (i.e., vanadium, molybdenum, and chromium), the beta phase can remain stable at room temperature, and mixtures of these phases are possible. Harder martensitic microstructures can form under rapid cooling conditions, often requiring postweld heat treatments. However, the weldability concern associated with martensite formation is minimal as compared to that associated with steels.

As mentioned, titanium alloys offer both excellent corrosion resistance and specific strength. Applications that rely on their corrosion resistance include heat exchangers, pressure vessels, waste storage, and tube and piping. They are resistant to stress corrosion cracking and perform well in marine environments. They are also biocompatible, which makes them a good choice for medical implants (Figure 12.11) such as hips, knees, and fasteners.

The widespread use of titanium alloys in the aerospace industry is driven almost exclusively by their high specific strength. In particular, high-performance military aircraft (Figure 12.12) rely

Figure 12.12 Titanium is used extensively in many military aircraft. (*Source:* Lockheed Martin).

heavily on titanium alloys for both air frames and skins. Their high specific strength can also create a performance advantage when they are used in sports equipment, such as bicycle frames, tennis rackets, and golf clubs.

Titanium alloys are weldable with most processes (except Shielded Metal Arc Welding) if proper precautions are taken. Two of the biggest welding concerns are interstitial embrittlement (or contamination cracking) and excessive HAZ and weld metal grain size. Titanium readily absorbs and dissolves interstitial elements such as oxygen, nitrogen, and hydrogen. In small amounts, these elements strengthen titanium quite profoundly. For example, the difference in tensile strengths among the CP titanium grades is mainly due to slight variations in oxygen and nitrogen content.

However, at larger amounts of absorption, significant embrittlement occurs. Embrittlement can occur quite rapidly above 500°C (930°F), which means that during welding, even regions on the part that are relatively far from the fusion zone are susceptible. Since the ambient atmosphere contains all the embrittling elements, protection (shielding) during welding is especially important. Therefore, when arc welding titanium it is common to use more thorough methods of shielding such as the use of larger nozzles, trailing shields, back-side shielding or purging, and in some cases welding in a glove box purged with an inert gas such as argon. It is also very important to properly remove oxides and contaminants prior to welding, and to use gasses of high purity.

The degree of contamination is often determined by the color of the oxide that forms during welding (Figure 12.13). The oxide color is a function of its thickness. A light silver or bronze-colored oxide indicates a mild contamination, while blue or purple oxides are an indication of more extensive atmospheric contamination and may result in the rejection of the weld based on the suspicion of embrittlement. A white, flaky oxide indicates severe contamination. However, it is important to remember that the oxide can be easily removed by wire brushing following welding, so completely relying on the oxide color as a measure of weld quality may not be the best approach.

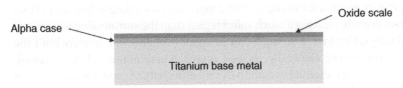

Figure 12.13 Colored oxide scale and/or alpha case formation is indicative of improper shielding when welding titanium.

A layer known as alpha case (Figure 12.13) may also form if shielding is not sufficient. Alpha case is a layer of titanium that has been highly enriched in alpha-stabilizing elements such as oxygen and nitrogen. It is brittle and susceptible to forming small cracks that can significantly reduce mechanical properties such as fatigue.

Titanium alloys solidify as a single-phase BCC beta structure and remain that way for some time during cooling. Since there are no other phases present to retard grain growth, weld metal and HAZ grain sizes can become quite large, resulting in significant reductions in ductility. The best approach to minimizing grain sizes is by reducing weld heat input, so utilizing extremely low heat input welding processes such as Electron Beam and Laser Welding can help keep grain sizes relatively small. Electron Beam Welding in particular is an ideal process for titanium since welding in a vacuum essentially eliminates any risk of interstitial embrittlement. In the right application,

Diffusion Welding is also a very good process for welding titanium. This is because titanium readily dissolves its own oxide and exhibits low creep strengths at elevated temperatures, two traits that are ideal for the Diffusion Welding process.

12.4 Copper Alloys

Copper alloys are used in a variety of engineering applications but are typically selected because of their electrical and thermal conductivity as well as corrosion resistance. Except for alloys containing beryllium (these alloys are strengthened by precipitation hardening), copper alloys are strengthened by solid solution and cold work, and in general have low to moderate strength but good ductility and toughness. Their microstructures are either single phase FCC or dual phase FCC + BCC. The family of copper alloys includes brass, a copper alloy containing zinc and possibly some tin or lead, and bronze, which is copper alloyed with tin, aluminum, or silicon.

Their resistance to saltwater corrosion makes them a good choice for marine applications such as tubing, boilers, and decorative hardware on boats and ships. Other applications that make use of their excellent corrosion resistance range from storage containers for nuclear fuel to caskets. They can be easily fabricated into a variety of shapes and can be readily cast. For example, early cannons were produced from copper alloy castings.

Copper alloys are also used for a variety of architectural and artistic purposes. The copper oxide that forms in the atmosphere can take on a variety of shades (depending on alloying elements added) from gold to red to brown, and thus is very versatile for creating different artistic effects. Under normal atmospheric exposure many of the Cu alloys will develop a greenish oxide, such as that seen over time on copper-roofed buildings. The physical properties of these alloys also make them popular in the musical instrument market, and for the production of bells and chimes. Most of the famous bells worldwide are cast from brass or bronze alloys.

Since cold work is a primary method for strengthening copper alloys, welding will result in HAZ recrystallization and grain growth, creating a much softer region than the surrounding base metal. In heavily cold worked alloys, this loss of strength can be significant and will severely limit the load-bearing capacity of the structure. Generally, the best approach to minimizing HAZ softening is to reduce heat input, but the high thermal conductivity and diffusivity of copper results in such rapid heat extraction during welding that high heat input is usually necessary. In fact, to achieve sufficient penetration with many alloys, weld preheating often becomes a requirement, especially as part thicknesses increase above about 0.25 in. (Figure 12.14). The use of helium-shielding gas can also significantly improve weld penetration. An example of the effect of preheat and shielding gas type on depth of fusion is shown in Figure 12.15.

Alloys that contain Zn, Cd, or P may be susceptible to porosity formation. This can usually be controlled by using a porosity resistant filler metal but can be problematic when welding autogenously. Removal of the oxide prior to welding by chemical or mechanical means is recommended. Most copper alloys are considered resistant to solidification and liquation cracking, even though they solidify as a single-phase FCC structure. Alloys containing tin and nickel may be susceptible to cracking under high restraint conditions because both elements tend to widen the solidification temperature range. Elements such as selenium, sulfur, and lead are often added to improve machinability, but they also increase susceptibility to solidification cracking, and therefore, should be avoided for applications that involve welding.

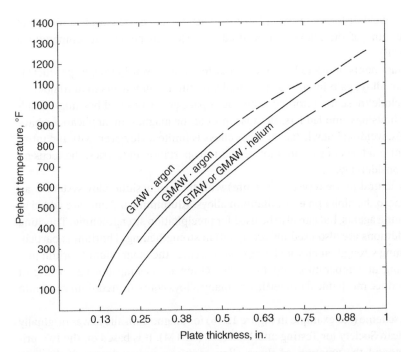

Figure 12.14 Typical preheat requirements for copper alloys as a function of thickness. (*Source:* Reproduced by permission of American Welding Society, ©*Welding Handbook*).

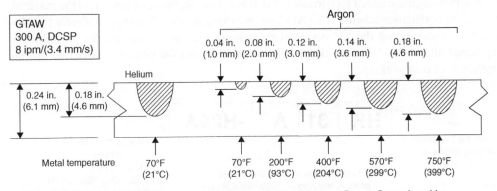

Figure 12.15 Effect of shielding gas and preheat on depth of fusion. (*Source:* Reproduced by permission of American Welding Society, ©*Welding Handbook*).

12.5 Magnesium Alloys

Magnesium alloys offer advantages similar to aluminum alloys but are even lower in density (lighter weight). They can be strengthened by heat treatment to achieve excellent strength-to-weight ratios (specific strength). The interest in magnesium is growing rapidly in the transportation industry where the drive toward more fuel-efficient (and electric) vehicles encourages reduced weight. Like aluminum alloys, these materials have a high CTE and wide solidification range making them susceptible to solidification cracking. The large CTE means they are susceptible to significant weld distortion as well. Most magnesium alloys are designed for use at room temperature, but some of the alloys can be used for extended periods at temperatures approaching 350°C (660°F).

Many magnesium components in use today are cast, but the trend toward wrought products is increasing as magnesium alloys gain interest as potential structural components in automobiles. Current applications include those for automotive and aerospace gear cases and housings, handheld tool and computer housings, and ladders. Because most of the magnesium applications have historically been castings, welding knowledge on these alloys is limited. However, with the introduction of more wrought products in the automotive and transportation industries, the focus on welding has increased considerably.

The primary microstructural phase (crystal structure) with the magnesium alloy system is an HCP alpha phase, similar to the alpha phase in titanium alloys. Aluminum and zinc are added as solid solution strengthening agents, but can also be used for precipitation strengthening. Thorium, silver, and rare earth additions are also used for precipitation strengthening. Thorium is particularly useful for maintaining strength at elevated temperatures since the magnesium-thorium precipitates resist coarsening at temperatures up to 350°C. Grain coarsening during casting of magnesium alloys can reduce mechanical properties, so many alloys contain zirconium as a grain refining agent.

The alloy designation scheme (see example in Figure 12.16) for magnesium alloys was originally developed by the American Society for Testing and Materials (ASTM). It is based on the two primary alloying additions, and the amounts of those alloy additions to the closest single digit. Common alloying additions include aluminum, manganese, zinc, and zirconium, but as the table shows, there are many more. The wrought alloys are based mainly on the Mg-Al-Zn and Mg-Zr-Zn systems with manganese added for corrosion resistance. The condition, or temper, of the material uses the same designation scheme as for Al alloys. For example, the "O" temper indicates that the material is annealed, and the "T" designations indicate the aging condition for precipitation strengthened alloys. The "H" condition is only associated with the wrought alloys that can be strengthened by cold working.

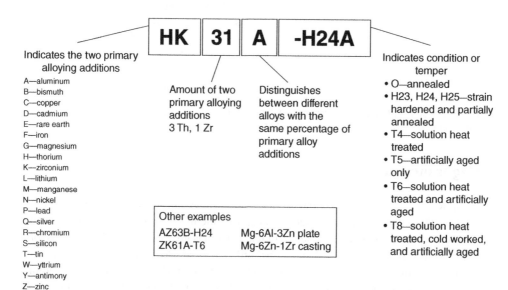

Figure 12.16 Alloy designation scheme for magnesium alloys.

Table 12.2 Comparison of selected mechanical and physical properties of aluminum and magnesium.

Property	Magnesium			Aluminum		
	Mg	AZ91D	HK31A-H24	Al	5454-H34	6061-T6
Density (g/cm^3)	1.738			2.71	2.68	2.70
Melting point (°C)	650			660	646	652
Volume change during solidification (%)	−4.2			−6.5		
CTE (μm/m · K)	25.2			23.6	23.7	23.6
Thermal conductivity (W/m · K)	167		238			
Yield strength (MPa)		150	200		240	275
Specific strength (YS/density)		86		115	39	102

AZ91D is a common die casting alloy (Mg-9Al-0.7Zn).
HK31A-H24 is a wrought alloy, Mg-3Th-0.6Zr, strain hardened and partially annealed (H24).

The physical and mechanical properties of two wrought magnesium alloys are presented and compared to aluminum in Table 12.2. Note the significant reduction in density as compared to aluminum. Even though magnesium alloys do not achieve the same yield strength levels as do aluminum alloys, due to their low density they quite often offer equal or greater specific strengths.

In addition to solidification cracking and weld distortion, other weldability problems associated with magnesium alloys include excessive grain size and porosity. Because magnesium alloys are single phase at elevated temperature, grain growth can be quite rapid resulting in the degradation of mechanical properties. The susceptibility to porosity is due to the high amounts of hydrogen in the alloys.

12.6 Test Your Knowledge

I) Fundamental Concepts—True/False

The following true/false questions pertain to some of the most important fundamental concepts in this chapter:

1) Aluminum alloy weld HAZ properties can be degraded from base metal properties by as much as 50%.
2) Aluminum alloys that are strengthened by cold work are known to be the most weldable of this alloy family.
3) A common approach to avoiding solidification cracking with aluminum alloys is to use a filler metal that promotes the formation of eutectic liquid.
4) Strain-age cracking associated with welding nickel-based alloys is known to occur as the weld is being made.
5) Strain-age cracking can be avoided by choosing nickel-based alloys with higher amounts of titanium and aluminum.
6) Titanium alloys require special shielding considerations when arc welding due to their sensitivity to interstitial embrittlement.
7) Electron Beam Welding is a poor choice for welding titanium.
8) Magnesium alloys are known to be susceptible to weld solidification cracking.
9) Copper alloys often require weld preheating because of their high thermal conductivity.

II) Solve a Welding Engineering Problem

The following challenge represents a typical problem Welding Engineers might encounter in their career:

Your company has just accepted a new project that involves welding a nickel-based alloy known as a superalloy. During the initial welding trials, cracking in the HAZ was observed following a required postweld heat treatment, and you have been assigned to figure out why and what to do about it. The first thing you do is check the chemistry of this alloy, and you learn it contains about 5 wt.% aluminum and 6 wt.% titanium. This immediately gives you a clue regarding why the cracking is occurring. Please explain the cracking mechanism and your plan to remedy the problem.

Recommended Reading for Further Information

ASM Handbook, Tenth Edition, Volume 6—"Welding, Brazing, and Soldering," *ASM International*, 1993.

Welding Metallurgy and Weldability of Nickel-Base Alloys, John Wiley & Sons, Inc., 2009.

13

Weld Quality

13.1 Weld Discontinuities and Defects

Weld quality can be quantitative or qualitative but is often a relative concept depending on the mechanical property requirements of the application. When considering weld quality, the "fitness-for-service" concept usually applies. A "fitness-for-service" approach is one that focuses on the intended service of the weldment or fabrication. For example, a weld flaw (imperfection) or discontinuity in a critical application such as a pressure vessel might be a rejectable defect per the applicable code, whereas the same imperfection might be acceptable for a weld used to fabricate shelving. Therefore, a discontinuity or flaw can be loosely defined as any type of perceived weld imperfection, and it only becomes a defect when the applicable welding code dictates that it is. More specifically, a discontinuity becomes a code-defined defect when it is expected to adversely affect the mechanical properties that are required for the given application.

In some cases, the distinction between a discontinuity and a defect requires a quantifiable measurement. For example, a code may specify that weld undercut up to 1/32 in. deep is an acceptable discontinuity, whereas undercut greater than 1/32 in. is a rejectable defect. On the other hand, surface cracks are almost always considered defects regardless of their size because they usually result in a significant degradation of mechanical properties. It is also important to point out that placing greater requirements on weld quality than is necessary for the application unnecessarily drives up the cost. Using the previous example, a section of shelving that was welded per the code requirements for a pressure vessel would likely be so expensive that few could afford it!

History has shown that weld defects can produce catastrophic results if they are not identified and not properly addressed. In the early 1940s, three World War II Liberty ships broke in half without warning due to suspected weld defects. Another famous and catastrophic failure attributed to a welding defect was the Alexander Kielland Norwegian offshore oil rig, which capsized in 1980 killing over 100 people.

Figure 13.1 shows a sampling of common arc weld discontinuities. Porosity (1B) consists of rounded pores that form in the weld for a variety of reasons, but typically because of shielding gas problems and/or the joint was not properly prepared prior to welding. Incomplete fusion (3) occurs when the weld metal does not fuse to the base metal, usually due to improper welding technique or procedures. Incomplete penetration (4) refers to a weld that did not completely penetrate through the joint, although this is not always a discontinuity since some joints do not require full

Welding Engineering: An Introduction, Second Edition. David H. Phillips.
© 2023 John Wiley & Sons, Inc. Published 2023 by John Wiley & Sons, Inc.
Companion Website: www.wiley.com/go/Phillips/WeldingEngineeringIntroduction

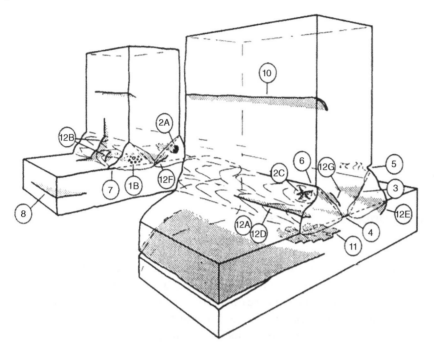

Figure 13.1 A sampling of common arc weld discontinuities. (*Source:* Reproduced by permission of American Welding Society, ©*Welding Handbook*).

penetration. Undercut (5) is a discontinuity at the toe of the weld caused by improper welding technique and characterized by a region of base metal that melted and was drawn into the weld metal creating a sharp groove. Overlap (7) is essentially a region of lack of fusion at the toe of the weld. Lamellar cracking (11) can occur near the weld when poor-quality steel is used. This type of cracking is much less common than it used to be many years back when practices for steel making were less advanced than they are today. The various discontinuities identified with a 12 and a letter refer to different forms of weld metal cracking. Two common forms of cracking—solidification and hydrogen cracking—were discussed previously in the welding metallurgy chapters.

As mentioned, the reason for the concern with the various types of weld discontinuities is the impact they may have on mechanical properties. There are many mechanical properties of special importance to weldments, depending of course on the intended service conditions. As discussed in Chapter 7, the most important properties are tensile strength, ductility, impact and fracture toughness, and fatigue strength. Any of these properties may be degraded by a welding discontinuity. In the following section, approaches to weld testing of these properties are reviewed. Many of these test methods play important roles in the qualification requirements for both welders and welding procedures.

13.2 Mechanical Testing of Weldments

13.2.1 Tensile Testing

Tensile strength is the maximum stress a material can withstand while under tensile loading. When evaluating the tensile strength of welds, it is common to test for both the strength of the weld metal itself as well as the strength of the joint. Figure 13.2 shows the methods for testing each.

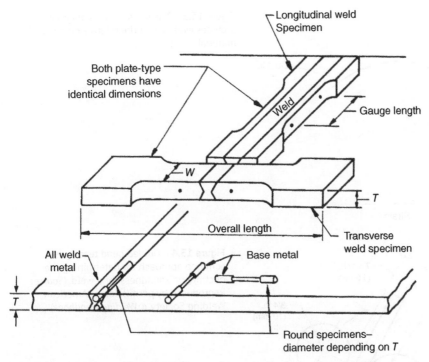

Figure 13.2 Weld metal and reduced section tensile test. (*Source:* Reproduced by permission of American Welding Society, ©*Welding Handbook*).

The weld metal tensile test requires that a test sample be machined entirely out of weld metal, whereas the entire joint can be tested for tensile strength using what is known as a transverse tensile coupon. This type of test will fail in the weakest part of the specimen which is often in the weld heat-affected zone. Therefore, while this test will identify the weakest region in a weldment, it usually provides no information regarding the weld metal itself.

13.2.2 Ductility Testing

Ductility can be loosely defined as the ability of a material to deform plastically without failure, and more specifically defined as % elongation of the sample during a tensile test. A stress–strain diagram (Figure 13.3) produced during a tensile test can be used to assess a material's ductility. A ductile material will undergo considerable plastic deformation (the nonlinear portion of the curve) before failure, while a brittle material will fail with little or no plastic deformation. Welding can result in significant loss of ductility, especially when welding transformation hardenable steels. Weld discontinuities such as slag inclusions can also degrade ductility.

Guided Bend testing (Figure 13.4) is a common approach to test a weld's ductility (or soundness) because it is a relatively easy, quick, and inexpensive test. The two main types of bend tests are transverse and longitudinal to the weld. Transverse tests focus on the root, face, or side of the weld (Figure 13.5), while longitudinal tests are tests that focus on either the root or face. Following the bend test, the surface of the test specimen is inspected for any discontinuities or cracks that may have formed during the test. Welding codes such as AWS D1.1 may allow certain discontinuities or groups of discontinuities that are less than a specified dimension, while rejecting those that are greater.

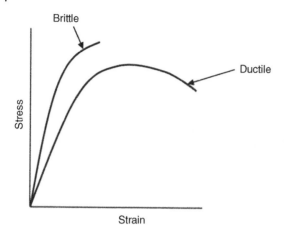

Figure 13.3 The stress–strain diagram provides evidence of the relative ductility of a material.

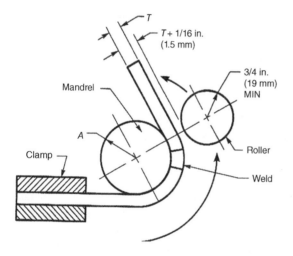

Figure 13.4 Guided bend testing is a common approach for determining the ductility or "soundness" of a weld. (*Source:* Reproduced by permission of American Welding Society, ©*Welding Handbook*).

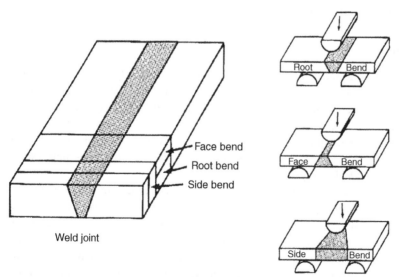

Figure 13.5 Transverse bend testing of a welded test coupon. (*Source:* Reproduced by permission of American Welding Society, ©*Welding Handbook*).

13.2.3 Toughness Testing

Toughness can be defined as the ability of a material to absorb energy and deform plastically before failure. A material that has good toughness has a combination of good tensile strength and good ductility, so the area under the stress–strain curve (Figure 13.6) is one way to represent a material's toughness. Some metals such as carbon steels exhibit good toughness above room temperature, but very poor toughness at lower temperatures. The toughness exhibited by a material can be significantly affected by the strain rate of the test, and the presence of a notch.

The most common type of toughness test for weldments is the Charpy V-Notch Impact Toughness Test. It measures impact toughness in the presence of a notch, although it is not

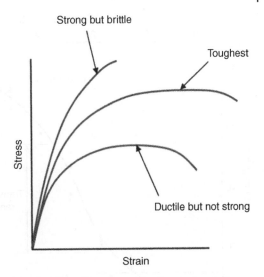

Figure 13.6 The area under the stress–strain curve is indicative of toughness.

actually considered a true fracture toughness test (discussed next). The Charpy test uses a weighted hammer on a pendulum that slams into the back of the notched specimen (Figure 13.7) breaking it. The amount the hammer continues to swing after breaking the sample is then measured using a scale and pointer on the testing apparatus. A tougher material will absorb more of the energy from the hammer causing it to swing less after impact as compared to a brittle material. This test is commonly used to assess the ductile-to-brittle behavior (Figure 13.8) typical with many carbon steels. Testing is therefore often conducted at very low temperatures. The Charpy V-Notch test is simple, fast, and inexpensive. Test results are considered to be more qualitative than quantitative, and therefore, mainly used for comparative purposes.

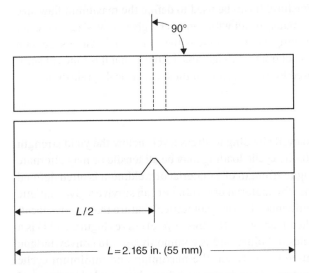

Figure 13.7 Charpy V-Notch impact toughness test specimen. (*Source:* Reproduced by permission of American Welding Society, ©*Welding Handbook*).

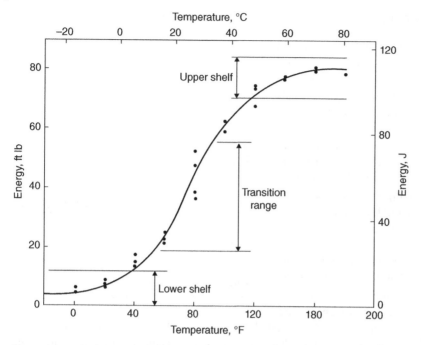

Figure 13.8 Typical ductile-to-brittle transition behavior of carbon steel as measured by the charpy impact test. (*Source:* Reproduced by permission of American Welding Society, ©*Welding Handbook*).

True fracture toughness is the measure of the stress necessary to propagate a preexisting crack. As compared to Charpy Impact testing, true fracture toughness testing uses low strain rates, and the test results can be used quantitatively to predict the load-bearing capabilities of materials that contain a flaw or a crack. Fracture toughness testing can be used for design criteria through an approach known as Linear Elastic Fracture Mechanics. This approach considers the flaw size, the component geometry, loading conditions, and fracture toughness of the material to assess the ability of a part with existing flaws to resist fracture. It can be used to define the maximum flaw size for a given stress or the maximum stress a material can withstand per a given flaw size. There are many approaches to fracture toughness testing. One example is the Compact Tension specimen shown in Figure 13.9. A common feature of all fracture toughness testing is that it is time consuming and expensive, so it is typically only used for weld testing in the most critical applications.

13.2.4 Fatigue Testing

Fatigue failure is a type of fracture due to cyclic loading at stress levels below the yield strength of the material. As shown in Figure 13.10, the cyclic loading may be all tensile or may alternate between tensile and compressive. The fatigue strength of a material or weldment is often defined as the number of cycles (or amount of time) the material or weldment can survive a given fatigue stress range prior to failure. It is highly influenced by any geometrical feature that creates localized stress concentrations during an applied load. An S-N (stress-cycle) curve (Figure 13.11) is a plot of stress range versus number of cycles to failure and is commonly used to convey fatigue test results. Stress range refers to the difference between the maximum and minimum cyclic stress. As this S-N curve shows, rolled beams perform better in fatigue than welded beams, and welded beams perform better than end welded cover plates.

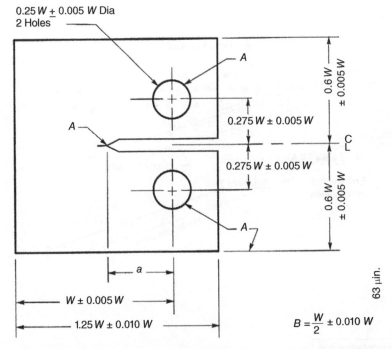

Figure 13.9 Compact tension fracture toughness test specimen. (*Source:* Reproduced by permission of American Welding Society, ©*Welding Handbook*).

The fatigue life of nonwelded material is generally divided into two phases: crack initiation and crack growth. Initially, the cyclic nature of the stress forms a crack. This phase may consume as much as 90% of what is the entire fatigue life of a part. Once the crack is initiated, stresses concentrate at the crack tip, causing the crack to propagate. Crack propagation may continue for some period of time prior to failure, which often occurs suddenly. Structural designs that include sharp corners or notches are prone to fatigue failure because the corners and notches act as stress concentrators which promote fatigue cracks.

In weldments, common regions of stress concentrations, and therefore likely fatigue crack initiation sites, include weld toes, weld roots,

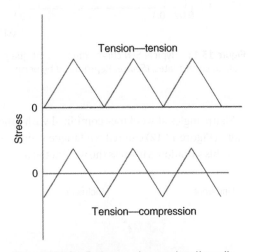

Figure 13.10 Fatigue cycles may be all tensile or tensile and compression.

and discontinuities. Such features may result in virtually no fatigue crack initiation phase, resulting in weldments exhibiting as low as 10% of the fatigue life of a nonwelded material. In a typical fatigue crack failure, crack propagation by fatigue continues until the remaining material can no longer support the load resulting in an overload failure. Therefore, fatigue failure fracture surfaces typically reveal evidence of fatigue crack growth combined with ductile dimple features, which provide evidence of the overload failure at the end.

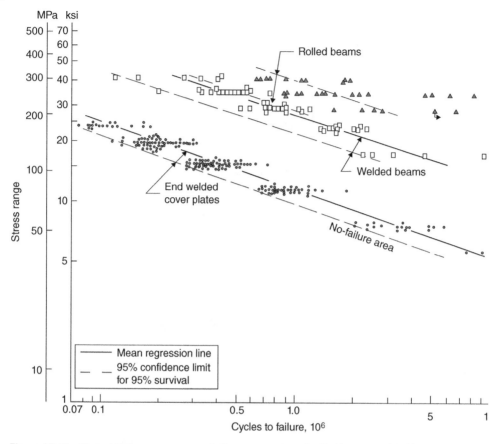

Figure 13.11 Typical S-N curve compares fatigue properties of rolled beams, welded beams, and end welded cover plates. (*Source:* Reproduced by permission of American Welding Society, ©*Welding Handbook*).

Sharp angles at weld toes combined with other discontinuities such as undercut and slag inclusions (Figure 13.12) can reduce fatigue properties even further. A common approach used to mitigate this situation is to dress the weld at the toe with a grinding operation to remove the discontinuity and/or reduce the toe angle. An additional weld pass may also be used to reduce toe angles or smoothen sharp transitions.

The design of the joint can also play a major role in fatigue life. For example, a butt joint will more evenly distribute a load, and therefore, would be expected to perform better under fatigue loading conditions than a lap joint. The AWS D1.1 Structural Welding Code for Steel includes fatigue design guidelines that address the design of the joint. These guidelines place various joint designs

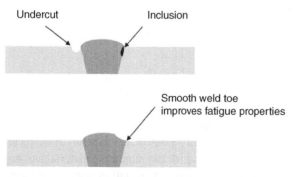

Figure 13.12 Weld discontinuities and sharp toe angles can significantly reduce fatigue life, but a simple grinding operation can improve the situation.

into six broad categories, lettered A through F, which reflect the different levels of stress concentration that can be expected in each joint design category.

Typical joints from Category B and E are shown in Figure 13.13. The category E joint involves a lap joint that would be expected to create greater stress concentrations and lower fatigue life than the joint selected from Category B. The fatigue design criteria for weldments for the various joint design categories are illustrated on a characteristic S-N curve (Figure 13.14). The categories are represented by the lines shown on the plot, and any combination of stress range and number of cycles that exceeds the category line will be expected to result in a fatigue failure.

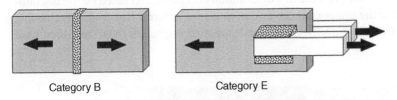

Category B   Category E

Figure 13.13   Two typical joint designs from category B and E.

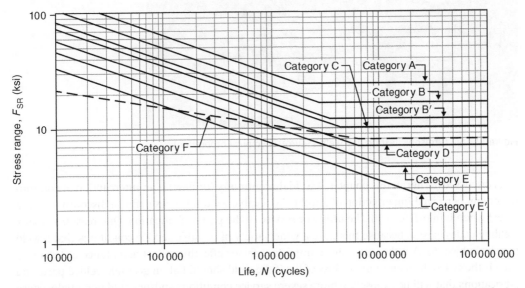

Figure 13.14   AWS D1.1 fatigue design guidelines reveal the expected fatigue performance of the various joint design categories. (*Source: AWS D1.1 2010, "Structural Welding Code–Steel"*).

This plot shows that a category B joint will perform better than a category E joint because it can survive a higher stress range and/or a given stress range for a longer time. These categories include factors of safety, and of course assume that the weldments are produced without defects. Interestingly, fatigue life is not dependent on mechanical properties such as tensile or yield strength, so changing to a higher strength base metal will not result in better fatigue properties.

13.3   Nondestructive Testing

Although valuable information is gained from conducting a mechanical test, the obvious disadvantage of this approach is it involves destroying the material or part being tested. To assess weld quality without destroying the part, Nondestructive Testing (NDT) techniques are often used. NDT refers to all possible methods for evaluating the performance of materials or fabrications without affecting their serviceability. The primary goal of all NDT methods is to predict and prevent failures through the detection of defects, usually under the guidance of the applicable code. NDT may be performed during fabrication/manufacturing and/or after exposure to service conditions. NDT measurements can also be used to predict the life of a component. For example, in-service inspection (Figure 13.15) can detect the presence of a discontinuity (or flaw), and based on the discontinuity size, the useful life of the component can be determined. The terms Nondestructive Testing (NDT), Nondestructive Evaluation (NDE), and Nondestructive Inspection (NDI) are often used interchangeably.

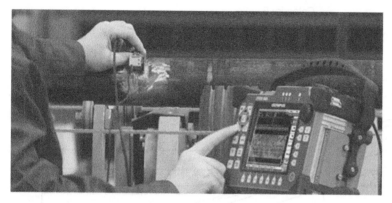

Figure 13.15   Ultrasonic testing of a weld. (*Source:* Olympus Corporation).

As discussed earlier in this chapter, a weld discontinuity (or flaw) is an interruption of continuity, or any perceived imperfection of a weld. Applicable welding codes define a discontinuity as a defect when it interferes with the utility or service of the part or fabrication. One reason so many welding codes exist is because there are a wide variety of possible service conditions that a weld may be subject to. Another even more important factor affecting defect acceptance criteria in a code is the risk of human injury or loss of life if the weld should fail. In general, welded parts and fabrications that will be exposed to more severe service conditions and/or could potentially cause human injury if they fail will be subject to the most severe acceptance criteria.

There are many types of NDT techniques that can be used for inspection of welded structures. These techniques vary in their simplicity, as well as their ability to detect different types and locations of discontinuities and defects. The most popular of these is the simple Visual Examination (VT). Most welds are visually inspected at some level, but Visual Examination, in addition to Liquid Penetrant Testing (PT) and Magnetic Particle Testing (MT), are limited to the detection of surface discontinuities, or in the case of MT, those that are very near the surface. Subsurface discontinuities are detected by a variety of techniques, with Radiographic Testing (RT), Ultrasonic Testing (UT), and Eddy Current Testing (ET) being the most popular. All approaches to NDT, including Visual Inspection, typically require personnel who are certified to conduct the test.

A nondestructive mechanical test, often called a proof test, may also be used to confirm that the welded structure or component will meet the service requirements. This is most often applied to pressure vessels or pipelines, where testing involves creating an internal pressure at some level above the operating pressure to ensure that catastrophic failure will not occur. All these techniques offer advantages and disadvantages, as described in the following pages.

13.3.1 Visual Examination

Visual Examination is the most widely used form of NDT because it is simple, fast, and economical. Although anyone knowledgeable about welding can conduct a Visual Examination, only an AWS Certified Welding Inspector (C.W.I.) is qualified to do so. Structural welds, particularly those made in the field, are almost always subject to a Visual Examination. As mentioned, this method can provide a lot of information about the weld quality but is limited to detecting surface discontinuities only. Although most surface discontinuities can be easily observed, it is not always obvious to the inspector that they should be categorized as rejectable defects, so Visual Examination can sometimes be subjective. Small surface cracks can sometimes be difficult to detect visually, so if there is a concern for cracking it is common to include other forms of NDT such as Liquid Penetrant Testing. Another very important purpose of a Visual Examination is to verify that proper fillet weld sizes are being made per the applicable print or procedure. This is a critical part of a visual inspection because an undersized

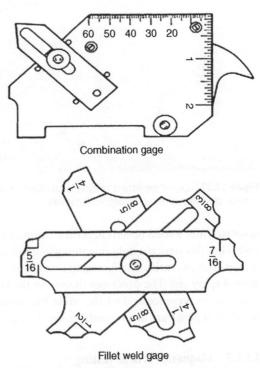

Combination gage

Fillet weld gage

Figure 13.16 Typical gauges used by an inspector during a visual inspection. (*Source:* Reproduced by permission of American Welding Society, ©*Welding Handbook*).

fillet weld will have lower strength than a weld made at the proper size. There are a variety of special gauges (Figure 13.16) and tools that are used by the inspector conducting the examination to measure discontinuity sizes, weld profiles, and weld sizes.

13.3.2 Liquid Penetrant Testing

Liquid Penetrant Testing is another simple, fast, portable, and inexpensive form of NDT. Other than lighting, it requires no external power and very little training. Like Visual Inspection, it is limited to surface discontinuities only. One big advantage that it offers over Visual Examination is that it is very good for detecting cracks (Figure 13.17).

Liquid Penetrant Testing procedures include many steps. The first step involves cleaning the part to remove oils, greases, dirt, and any other contaminants. After drying the cleaned surface, brightly colored penetrant is then applied. This is typically done with an aerosol can much like

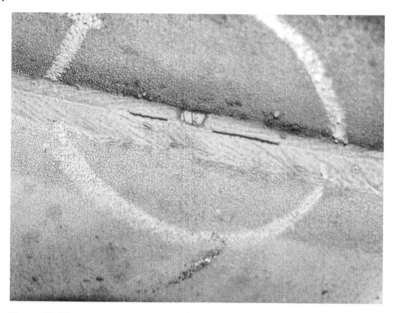

Figure 13.17 Liquid penetrant testing is an effective and simple method for revealing cracks on the surface of a weld. (*Source: Funtay/AdobeStock*).

spray painting. Excess penetrant is then removed from the surface of the part, usually with a cloth. Following this step, the only remaining penetrant will be contained within any surface discontinuities such as cracks. The final step involves the application of a white developer, also using an aerosol spray can. The developer draws out the colored penetrant from the discontinuity clearly revealing its presence against the white background of the developer. Another approach uses a fluorescent penetrant that is easily revealed using special lighting.

13.3.3 Magnetic Particle Testing

Magnetic Particle Testing (Figure 13.18) is based on the principle that magnetic lines of force in a ferromagnetic material are disrupted by changes in the continuity of the material. This method is

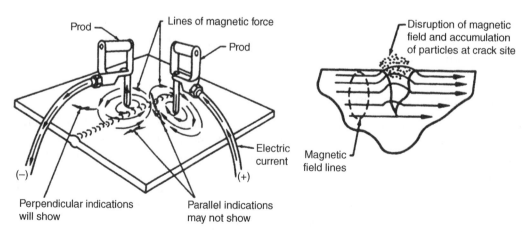

Figure 13.18 Basic principles of magnetic particle testing. (*Source:* Reproduced by permission of American Welding Society, ©*Welding Handbook*).

therefore limited to ferromagnetic materials (iron and steel) since the part must be magnetized to generate the lines of flux. Any discontinuities that are present in the magnetic field will cause a disruption in the magnetic flux producing "flux leakage" around the discontinuity. To detect this leakage, a powder consisting of very fine colored iron oxide particles is applied to the surface of the part. These particles are attracted to the magnetic field disruption causing them to accumulate at the discontinuity and reveal its presence. The powder can either be in dry form or suspended in a liquid bath (normally a light petroleum distillate).

This method is applicable to both surface discontinuities, and those that may lie just below the surface. It is especially effective at detecting cracks and is relatively fast, inexpensive, and portable, but does require electrical power. Minimal training is required. Parts to be inspected must be cleaned before and after testing. The magnetic field can be applied either by placing the part in a magnetic field or by passing a current through the material to induce a field. Both AC and DC current can be used to induce magnetic fields. One disadvantage of Magnetic Particle Testing is that the discontinuity may only be detected if it lies perpendicular to the lines of flux (Figure 13.19). This is because cracks or discontinuities that are oriented parallel to the lines of flux result in little or no disruption to the magnetic field.

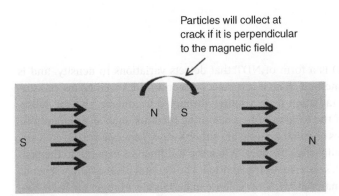

Particles will collect at crack if it is perpendicular to the magnetic field

Figure 13.19 With magnetic particle testing, discontinuities may only be detected if they are perpendicular to the lines of flux.

There are many different approaches to this technique that allow for testing of a large range of part sizes and flaw orientations. A common approach known as Circular Magnetization produces circular magnetic flux lines around the part. This approach is good for detecting flaws that are parallel to the length of the part. Longitudinal Magnetization produces a magnetic field aligned in the direction of the part to detect flaws that Circular Magnetization might miss. For large parts, a local electric field can be applied to produce a magnetic field using electric "prods" (shown in Figure 13.18). These prods must be positioned in multiple locations to vary the lines of flux around the weld to ensure that all discontinuities are detected. This technique may result in some remnant magnetization of the part, which may be undesirable in certain applications. The yoke method (Figure 13.20) is also good for large parts and delivers a magnetic field to the part without passing current. It therefore offers the advantage of no arcing to the part that can sometimes be a problem when using the prods.

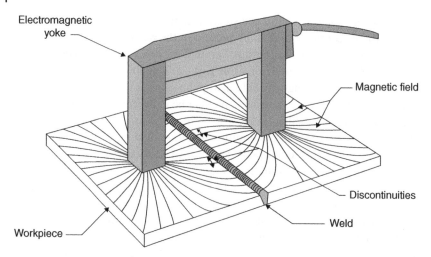

Figure 13.20 The yoke method works well for large parts, and offers the advantage of no arcing to the part. (*Source:* Welding Essentials, Second Edition).

13.3.4 Radiographic Testing

Radiographic Testing (Figure 13.21) is a form of NDT that detects variations in density, and is known as a volumetric inspection method since it can detect discontinuities well below the surface. It relies on the penetration of radiation from a source through a weldment, and the subsequent exposure on the other side of the weldment to a recording medium such as photographic film or a fluorescent screen. The types of radiation used are either X-rays (generated using an X-ray machine) or gamma rays (from a radioactive isotope). When the weldment is exposed to the penetrating radiation, it can be absorbed, scattered, or transmitted. The amount of absorption depends on the density and thickness of the material, with more dense materials absorbing more radiation. When more radiation is absorbed relative to the rest of the material being tested, less of it passes through to the recording medium. Regions of less exposure on the recording medium will show up as white or lighter shades of gray relative to the regions of greater exposure.

Radiographic Testing can be used with virtually all materials. It is widely used in applications where the detection of subsurface flaws or discontinuities is required, although sensitivity is reduced as part thicknesses increase. The phrase "X-ray quality weld" is often used to describe welds that are considered to be the most critical, and therefore subject to the most thorough forms of testing. For example, nuclear pressure vessels and pipelines are usually inspected using this technique to ensure that there are no subsurface defects that could lead to catastrophic failure.

Examples of exposed film (radiographs) after testing are shown at the top of Figure 13.22, and the accompanying metallographic sample containing the weld discontinuity (in this case a slag inclusion) is shown below it. In addition to slag inclusions, porosity is also easily detected by this method because both discontinuities represent regions of reduced density, and therefore, reduced absorption of radiation. Tungsten inclusions, which can occur when welding with the Gas Tungsten Arc Welding process, are also easy to detect because they are denser than the surrounding material. As a result, they absorb more radiation producing a bright spot on the film. However, cracks are not always easy to detect with Radiographic Testing, especially if they are relatively small or not aligned properly with the transmitted radiation.

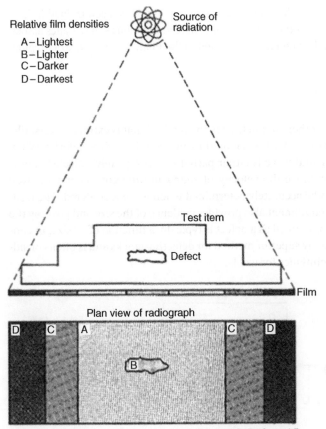

Note: Relative film densities from light to dark are A, B, C, and D.

Figure 13.21 Radiographic testing identifies flaws by detecting variations in density. Relative film densities from light to dark are A, B, C, D. (*Source:* Reproduced by permission of American Welding Society, ©*Welding Handbook*).

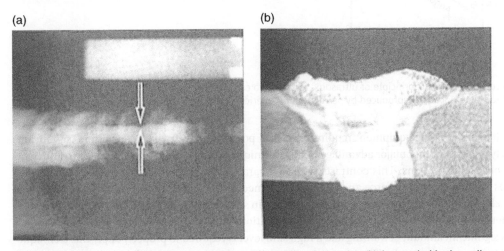

Figure 13.22 The slag inclusion seen in the metallographic cross-section (b). is revealed in the radiograph above (a). (*Source:* Reproduced with permission of American Welding Society).

Radiographic Testing requires highly skilled certified personnel who have been trained in how to apply the technique as well as interpret the radiographs. There are obvious safety issues associated with this form of testing, especially when it is conducted in the field. As a result, strict regulations govern its use.

13.3.5 Ultrasonic Testing

Ultrasonic Testing (Figure 13.23) is another volumetric testing method that is extremely versatile. It uses the principle of high frequency (1–10 MHz) sound transmission through a material. When a discontinuity is encountered, the sound wave is either partially or completely reflected, and is then analyzed by a receiving device. Since the velocity of sound in different materials is well known, a discontinuity's location can be accurately determined when it is encountered. The feedback signal on the CRT screen of the equipment also provides evidence of the size and possibly the type of discontinuity, and whether it is defined as a defect is typically a function of its size. In contrast to Radiographic Testing, it is usually superior method for detecting cracks and is often considered the most sensitive of all NDT techniques for welds.

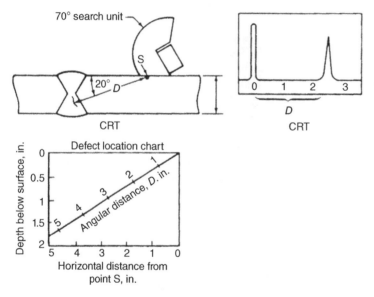

Figure 13.23 A basic principle of ultrasonic testing is to correlate feedback signals with discontinuity sizes and locations. (*Source:* Reproduced by permission of American Welding Society, ©*Welding Handbook*).

Ultrasonic Testing equipment requires electrical power but is very portable with no significant safety issues. Another major advantage of this technique is that access to only one side is sufficient to conduct inspections. This contrasts with Radiographic Testing which requires access to both sides (for the radiation source and the recording medium). For this reason, Ultrasonic Testing is widely used to inspect components that have been in service where access to both sides of the weld is not possible, which is often the case with fabrications such as pressure vessels and pipelines.

A significant amount of training and a high skill level is required to properly apply this technique and to interpret the results. The transducer that delivers the ultrasound and the collector (or search unit) must be appropriately placed to properly inspect the component, and correct interpretation is critical to avoid "false positives." A liquid couplant is needed between the transducer and the part. Rough or irregularly shaped surfaces may be difficult or impossible to inspect with this method.

There are many approaches to detecting discontinuity types, sizes, and locations with Ultrasonic Testing. Two approaches for determining the size of a discontinuity are shown in Figure 13.24. The approach at the top of the figure involves correlating the amplitude of the feedback signal ("blip") to the size of the discontinuity. In the other approach, the length of a discontinuity can be determined by moving the search unit along the plate until the signal is lost.

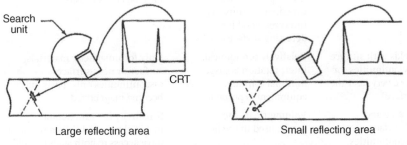

Large reflecting area   Small reflecting area

Strength of signal provides evidence of discontinuity size

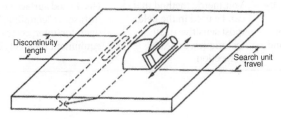

Transducer can be move longitudinally to measure discontinuity length

Figure 13.24   Two approaches to determine a discontinuity size with ultrasonic testing. (*Source: Reproduced by permission of American Welding Society, ©Welding Handbook*).

In summary, there are many NDT choices, each offering different capabilities, levels of training required, portability, and expense. Table 13.1 summarizes the common techniques described in this book, including typical applications, advantages, and limitations.

13.4   Introduction to Fractography

Fractography refers to a method for determining the cause of a failure by examining the microscopic features of the failed fracture surface. This technique can be quite effective for determining why welds fail. It typically requires the use of a powerful Scanning Electron Microscope (SEM) as well as a knowledge of the characteristic fracture surface features.

A significant amount of information about the material and failure mechanism can be revealed on a fracture surface. For example, Figure 13.25 shows fracture surface features that are characteristic of a ductile failure and are known as ductile dimples or shear dimples. This type of fracture surface

Table 13.1 A summary of common NDT techniques.

Technique	Applications	Advantages	Limitations
Visual	Welds with surface discontinuities	Economical, fast, minimal training required	Limited to surface discontinuities, difficult to detect cracks, somewhat subjective
Liquid penetrant	Welds with surface discontinuities	Use with all materials, inexpensive, minimal training, portable, expedient, results easy to interpret, good for detecting surface cracks	Proper surface prep required, limited to surface discontinuities, false positives possible
Magnetic particle	Welds with surface discontinuities (or those very near the surface)	Relatively economical, interpretation is easy, minimal training, equipment is portable	Only ferromagnetic materials, surface or near-surface discontinuities only, parts may become magnetized
Radiography	Surface and subsurface discontinuities, works with virtually all materials	Volumetric method that can be used in the field	Sensitivity is a function of material type and thickness, must have access to both sides of weld, interpretation can be difficult, skilled operator/interpreter required, safety issues, expensive
Ultrasonic	Can detect virtually all weld discontinuities in most materials and thicknesses	Volumetric method that can be used in the field, most sensitive of all techniques, one-side inspection	Need good surface conditions for ultrasonic "coupling," skilled operator/interpreter required, equipment relatively expensive

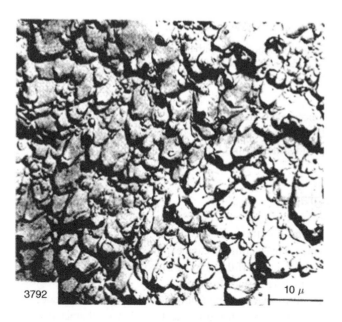

3792   10 μ

Figure 13.25 Ductile dimple fracture surfaces indicate a ductile or a simple overload failure.

indicates that the failure occurred due to overload, such as the case when a part is underdesigned, or the weld is too small for the load it is expected to support.

Fatigue cracks typically create very distinctive fracture surfaces. Figure 13.26 shows a typical fracture surface. Features known as beach marks and striations provide evidence of the cyclic progression of the crack tip during its growth. The very fine striations (Figure 13.26a) represent individual stress cycles, and generally are aligned perpendicular to the fatigue crack direction. The schematic on the right (Figure 13.26b) represents a typical fatigue fracture surface, which often also reveals beach marks that point to the crack origin, as well as a region of static or overload failure. The overload failure occurs after the fatigue crack gets so large that the remaining material can no longer support the load. This portion of the fracture surface would be expected to reveal the ductile dimple features shown in Figure 13.25.

A brittle fracture surface will exhibit what is known as chevron markings (Figure 13.27). A weld that fractures this way is known to have poor impact properties or notch toughness. For example,

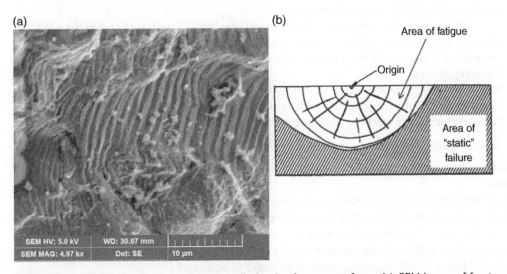

Figure 13.26 Fatigue failures produce very distinctive fracture surfaces. (a). SEM image of fracture surface (*Source:* Ontiveros et al. (2017). Thermodynamic entropy generation in the course of the fatigue crack initiation. Fatigue Fract Engng Mater Struct, 40: 423–434. Reproduced with permission from John Wiley & Sons.) and (b). Schematic representation of typical fatigue fracture surface.

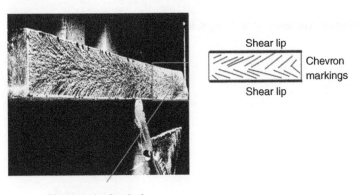

Figure 13.27 Chevron markings on a fracture surface indicate poor impact properties.

steels that are Charpy V-Notch tested at temperatures below their ductile-to-brittle transition temperature would be expected to exhibit this type of fracture surface. These examples represent a small sampling of the many characteristic fracture surface features that can be evaluated to determine the cause of a failure.

13.5   Test Your Knowledge

I) Fundamental Concepts—True/False

The following true/false questions pertain to some of the most important fundamental concepts in this chapter:

1) A weld discontinuity is always considered a defect.
2) The weld discontinuity known as undercut is a region of base metal that melted and was drawn into the weld metal creating a sharp groove.
3) The weld discontinuity known as overlap is essentially lack of fusion at the toe of the weld.
4) All welds should be subject to the same requirements, regardless of their intended service.
5) Assessment of the guided bend test for welds involves inspection of the surface of the sample following the test.
6) Welded structures usually exhibit about 80% the fatigue life of the unwelded base metal.
7) Visual Examination is an ideal way to detect surface cracks.
8) Radiographic Testing works by detecting variations in density.

II) Solve a Welding Engineering Problem

The following challenge represents a typical problem Welding Engineers might encounter in their career:

Your latest assignment as a Welding Engineer is to select an inspection method for a weldment between two very thick plates that require more than 50 passes to complete the weldment. Each pass is wire brushed to remove slag, so there is concern for the possibility of internal slag includes. The inspections will have to be conducted in the field, and you will only have access from one side of the weldment. Which inspection method do you choose, and why?

Recommended Reading for Further Information

AWS Welding Handbook, Ninth Edition, Volume 1—"Welding Science and Technology," American Welding Society, 2001.

14

Codes, Standards, and Welding Qualification

14.1   Introduction to Standards

Welding codes are very important to the Welding Engineer because they govern and guide welding activities to ensure safety, reliability, and quality of the applicable weldment or welded structure. Before beginning a discussion on welding codes, it is helpful to review a few definitions. AWS considers a "standard" to be any publication or document from the following six general categories: codes, specifications, recommended practices, classifications, methods, and guides. Codes (such as the common AWS D1.1, Figure 14.1) and specifications are similar in that they make frequent use of the word "shall" to indicate mandatory requirements. Their use is often a requirement per federal or state laws and regulations. For example, state transportation departments may require adherence to the AWS Bridge Welding Code (D1.5) for welding of bridge fabrications in that state. Codes and specifications are typically incorporated in the contract for producing a welded fabrication, and therefore, establish important information such as inspection methods and acceptance criteria for a weld, and methods for qualifying welders and weld procedures. The main difference between a code and a specification is that codes typically apply to processes (such as welding) and specifications typically apply to materials or products.

Recommended practices, classifications, methods, and guides are generally considered voluntary standards. They frequently make use of the words "should" and "may" (as opposed to "shall"), but they may become a requirement when invoked by a contract or applicable code or specification. Recommended practices typically describe general industry practice, whereas classifications, methods, and guides tend to provide more specific information regarding the best practical methods for performing a given task.

Standards originate from a wide variety of organizations, companies, technical societies, and government agencies. AWS (American Welding Society) and ASME (American Society of Mechanical Engineers) are two technical societies that provide the most used welding codes in the United States. Government agencies such as ANSI (American National Standards Institute) and international organizations (Figure 14.2) such as ISO (International Organization for Standardization) establish and publish numerous standards that are important to welding operations. Manufacturing companies and military agencies quite often develop and implement their own codes and specifications, although these are usually based on and are similar to technical society codes such as AWS.

Adding further to the complexity and the volume of available standards pertaining to welding is the fact that each industry sector or product type may require a different code. Table 14.1 shows the

Welding Engineering: An Introduction, Second Edition. David H. Phillips.
© 2023 John Wiley & Sons, Inc. Published 2023 by John Wiley & Sons, Inc.
Companion Website: www.wiley.com/go/Phillips/WeldingEngineeringIntroduction

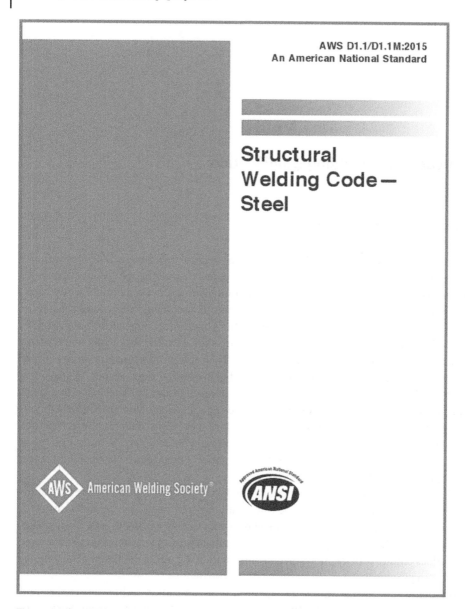

Figure 14.1 AWS D1.1, a very common welding code. *Photo credit:* American Welding Society.

Figure 14.2 ISO (International Organization for Standardization) is the world's largest developer of international standards. *Photo credit:* International Organization for Standardization (ISO).

Table 14.1 Examples of the wide range of welded product types and the various organizations that publish applicable welding standards.

Product areas covered by the standards issued by various organizations[a]

Product	AAR	AASHTO	ABS	AISC	API	AREMA	ASME	NBBPVI	ASTM	AWS	AWWA	FED	PFI	SAE	UL
Base metals		X	X		X	X	X		X	X	X	X	X		X
Bridges		X		X						X		X			
Buildings				X		X				X					
Construction equipment										X		X	X		
Cranes, hoists							X			X					
Elevators, escalators							X								
Filler metals			X				X			X		X	X		
Food, drag equipment							X								
Machine tools										X					
Military equipment												X			
Power-generation equipment			X				X	X				X			
Piping			X		X		X			X	X	X	X		
Presses										X					
Pressure vessels, boilers			X		X		X	X		X					
Railway equipment	X					X				X					
Sheet metal fabrication										X					
Ships			X							X		X			
Storage tanks				X	X					X	X				X
Structures, general						X				X					
Vehicles										X		X	X		

Source: Reproduced by permission of American Welding Society, ©*Welding Handbook.*
a) AAR, Association of American Railroads; AASHTO, American Association of State Highway and Transportation Officials; ABS, American Bureau of Shipping; AISC, American Institute of Steel Construction; API, American Petroleum Institute; AREMA, American Railway Engineering and Maintenance-of-Way Association; ASME, American Society of Mechanical Engineers; ASTM, American Society for Testing and Materials; AWWA, American Water Works Association; FED, US government standards; NBBPVI, National Board of Boiler and Pressure Vessel Inspectors/Uniform Boiler and Pressure Vessel Laws Society; PFI, Pipe Fabricators Institute; SAE, Society of Automotive Engineers; UL, Underwriters Laboratories.

wide range of possible product types that are involved in welding, and the numerous organizations that generate and publish applicable welding standards for that product or industry sector. As discussed previously, the primary reason for needing so many codes is the wide range in mechanical property requirements and safety factors associated with welded products and fabrications (Figure 14.3).

Figure 14.3 The various products and industry sectors involving welding have a wide range of requirements, safety factors, and applicable codes and specifications.

In addition to AWS and ASME, welding-related standards published by AISC (American Institute of Steel Construction) and API (American Petroleum Institute) are common in the United States. AISC (Figure 14.4) represents and serves the structural steel industry in the United States. It promotes structural steel fabrication through research, education, standardization, technical assistance, and quality control. It is well known in the steel fabrication industry for its "Manual of Steel Construction" which serves as a specification for the design, fabrication, and erection of buildings. It references AWS D1.1 for welding procedure and performance qualification.

API is a major national trade association that acts on behalf of the petroleum industry. It provides for standardization of engineering specifications pertaining to drilling and construction equipment including pipelines and storage tanks, and addresses design, fabrication, welding, inspection, and repair. It is well known for its API 1104 standard, "Welding of Pipelines and Related Facilities."

ASME (Figure 14.5) is responsible for over 600 technical standards, but the common standard associated with welding is known as the "Boiler and Pressure Vessel Code." This vast standard

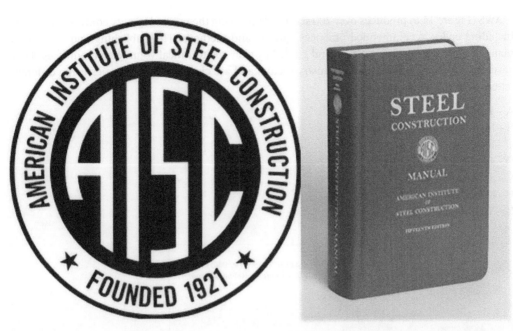

Figure 14.4 The AISC manual of steel construction is well known in the steel fabrication industry. *Photo credit:* American Institute of Steel Construction.

Figure 14.5 The ASME "Boiler and Pressure Vessel Code," B31.1, and B31.3 are important codes pertaining to welding in the nuclear industry. *Photo credit:* The American Society of Mechanical Engineers.

provides mandatory requirements and guidance for the construction of steam boilers and other pressure vessels. It is referenced in the safety regulations of most states and cities in the United States, as well as by many federal agencies. It is made up of 12 separate sections that range from rules for construction of power boilers, to NDE, to materials. Section IX establishes welding and brazing qualifications and is frequently used for the qualification of welders and welding specifications. The ASME B31.1 (Power Piping) and B31.3 (Process Piping) codes are quite commonly used when welding pipelines. The ASME welding standards govern and guide much of the most critical welding applications, such as welding in the nuclear industry.

AWS (Figure 14.6) produces over 200 standards covering the use and quality control of welding, including codes, recommended practices, guides, and methods. The publishing and revising of many of these documents is driven and guided by over 200 technical volunteer committees. They cover a wide range of topics and industry sectors and are classified by categories using different letters from the alphabet as follows:

A – Fundamentals
B – Inspection and Qualification
C – Processes
D – Industrial Applications
E – Safety and Health
F – Materials
G – Welding Equipment

Figure 14.6 The Bridge Welding Code is one example of over 200 standards published by AWS pertaining to welding. *Photo credit:* American Welding Society.

14.2 AWS D1.1–"Structural Welding Code–Steel"

The AWS D1.1 "Structural Welding Code—Steel" is one of the most widely used welding codes and will be discussed here in some detail. It is approved by ANSI and is considered an American National Standard. D1.1 contains the requirements for fabricating and erecting steel structures and is applicable to carbon and low-alloy steels with yield strengths equal to or less than 100 ksi, and thicknesses of 1/8 in. or greater. It does not apply to pressure vessels or pressure piping. It consists of eight clauses (like chapters) that cover topics ranging from design to qualification to inspection. The specific D1.1 clauses with a few examples of topics covered are as follows:

Clause 1. General requirements

Clause 2. Design of welded connections: common requirements of nontubular and tubular connections, specific requirements for nontubular connections (statically or cyclically loaded), specific requirements for cyclically loaded nontubular connections, specific requirements for tubular connections

Clause 3. Prequalification of Welding Procedure Specifications (WPSs): welding processes, base/filler metal combinations, limitation of variables

Clause 4. Qualification: welding procedure specification (WPS), type of qualification tests, performance qualification

Clause 5. Fabrication: preheat and interpass temperatures, backing, preparation of base metal, repairs, welding environment

Clause 6. Inspection: contractor responsibilities, acceptance criteria, nondestructive testing procedures

Clause 7. Stud welding: mechanical requirements, workmanship, technique, stud application qualification requirements, fabrication and verification inspection requirements

Clause 8. Strengthening and repairing existing structures: design for strengthening and repair, fatigue life enhancement, workmanship and techniques, quality

14.2.1 Welding and Welder Qualification

A few of these clauses will now be reviewed with emphasis on Clause 3 and Clause 4 that cover qualifications. Clause 3 addresses prequalification of WPSs, while Clause 4 describes qualification of procedures (that are not prequalified in Clause 3) and personnel. Clause 3 covers a wide range of prequalified welding processes, base and filler metal combinations, weld parameters and procedures (Table 14.2), and joint designs (Figure 14.7). Procedures from this section are considered prequalified, meaning, they are exempt from the qualification requirements of Clause 4. The use of prequalified procedures can save the fabricator significant amounts of money and time by avoiding the need to have to qualify a procedure. The disadvantage of prequalified procedures, however, is that there are limitations regarding the choice of welding processes and joint designs. Welders using a prequalified procedure must be qualified per the requirements of Clause 4.

Clause 4 guides the qualification of both welding procedures (those that are not prequalified) and welding personnel. Adherence to this section of the code ensures a level of welding quality required for structural steel fabrications. There are three very important records (or documents) that verify if welding quality is being maintained: the Procedure Qualification Record (PQR), the Welding Procedure Specification (WPS), and the Welder Performance Qualification (WPQ).

Table 14.2 Example of prequalified welding procedure requirements in AWS D1.1.

Variable	Position	Weld type	SMAW	SAW Single	Parallel	Multiple	GMAW/FCAW
Maximum electrode diameter	Flat	Fillet	5/16 in. (8.0 mm)				1/8 in. (3.2 mm)
		Groove	1/4 in. (6.4 mm)				
		Root pass	3/16 in. (4.8 mm)				
	Horizontal	Fillet	1/4 in. (6.4 mm)				1/8 in. (3.2 mm)
		Groove	3/16 in. (4.8 mm)	Requires WPS qualification test			
	Vertical	All	3/16 in. (4.8 mm)				3/32 in. (2.4 mm)
	Overhead	AN	3/16 in. (4.8 mm)				5/64 in. (2.0 mm)
Maximum current	All	Fillet	Within range of recommended operation by the filter metal manufacturer	1000 A	1200 A	Unlimited	Within the range of recommended operation by the filler metal manufacturer
	All	Groove weld root pass with opening		600 A	700 A		
		Groove weld root pass without opening			900 A		
		Groove weld fill passes			1200 A		
		Groove weld cap pass		Unlimited			

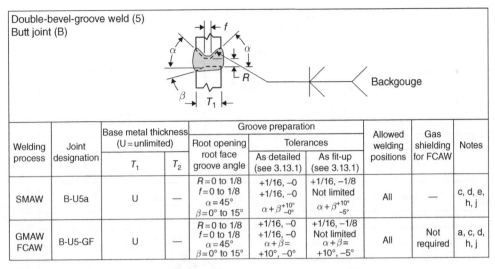

		Base metal thickness (U = unlimited)		Groove preparation			Allowed welding positions	Gas shielding for FCAW	Notes
Welding process	Joint designation	T_1	T_2	Root opening root face groove angle	Tolerances				
					As detailed (see 3.13.1)	As fit-up (see 3.13.1)			
SMAW	B-U5a	U	—	$R = 0$ to 1/8 $f = 0$ to 1/8 $\alpha = 45°$ $\beta = 0°$ to 15°	+1/16, −0 +1/16, −0 $\alpha + \beta^{+10°}_{-0°}$	+1/16, −1/8 Not limited $\alpha + \beta^{+10°}_{-5°}$	All	—	c, d, e, h, j
GMAW FCAW	B-U5-GF	U	—	$R = 0$ to 1/8 $f = 0$ to 1/8 $\alpha = 45°$ $\beta = 0°$ to 15°	+1/16, −0 +1/16, −0 $\alpha + \beta =$ +10°, −0°	+1/16, −1/8 Not limited $\alpha + \beta =$ +10°, −5°	All	Not required	a, c, d, h, j

Figure 14.7 Example of a prequalified joint geometry in Clause 3 of AWS D1.1.

Before describing the PQR, WPS, and WPQ, it is helpful to first review the general approach to welding procedure development. There are numerous steps involved in developing a welding procedure, but the most basic steps (with brief descriptions) are as follows:

1) **Determination of applicable code, base metal, and welding process**

   There are many aspects to determining the applicable code, but as mentioned previously, many times it will be mandated by a government agency. Contract documents also usually include code requirements for the intended application.

2) **Drafting of a preliminary welding procedure**

   Once the code is known and a process is chosen, the initial phases of developing a new welding procedure can begin. There are many sources that can be used to guide welding procedure development, including AWS Handbooks and the numerous AWS guides and recommended practices, power supply manufactures, and filler metal suppliers. An existing procedure may be used as a starting point as well, assuming of course it is for a similar application.

3) **Welding of test coupon(s)**

   Upon determining the preliminary procedure details, the test coupon(s) can be welded. The code will guide the details associated with this step, including the appropriate types of coupons and their dimensions, and locations of the test specimens to be removed from the welded coupon for testing.

4) **Mechanical and/or nondestructive testing of test coupons**

   Per the code, the required mechanical tests and/or nondestructive tests can then be conducted. In some cases, certain nondestructive tests can be used as substitutes for destructive tests.

5) **Preparation of WPS and PQR to include test results, welding parameters, and other process details**

   Once the test results are known, the PQR can be developed as well as the final WPS. These two important documents are linked to each other.

The PQR is a record of procedures, parameters, and mechanical and/or nondestructive test results that verify the preliminary WPS used to weld the test plates was acceptable. An important element of the PQR is the mechanical test details and test results that establish the strength, ductility, and overall soundness of the test weld. A portion of an AWS recommended PQR form is shown in Figure 14.8. The PQR must be established and approved by the manufacturer or contractor before the final WPS is developed and production welding begins.

Procedure Qualification Recod (PQR) # __________
Test Results

TENSILE TEST

Specimen No.	Width	Thickness	Area	Ultimate Tensile Load, lb	Ultimate Unit Stress, psi	Character of Failure and Location

GUIDED BEND TEST

Specimen No.	Type of Bend	Result	Remarks

VISUAL INSPECTION
Appearance__________
Undercut __________
Piping porosity__________
Convexity__________
Test date __________
Witnessed by__________

Radiographic-ultrasonic examination
RT report no. __________ Result __________
UT report no. __________ Result __________

FILLET WELD TEST RESULTS

Minimum size multiple pass Macroetch
1. ______ 3. ______
2. ______

Maximum size single pass Macroetch
1. ______ 3. ______
2. ______

Other Tests

All-weld-metal tension test
Tensile strength, psi __________
Yield point/strength, psi __________
Elongation in 2 in., % __________
Laboratory test no. __________

Welder's name __________ Clock no.__________ Stamp no. __________

Tests conducted by __________ Laboratory
Test number __________
Per __________

We, the undersigned, certify that the statements in this record are correct and that the test welds were prepared, welded, and tested in accordance with the requirements of section 4 of AWS D1.1, (__________) Structural Welding Code—Steel.
(year)

Signed __________
Manufacturer or Contractor
By__________
Title__________
Date __________

Form E-1 (Back)

Figure 14.8 Typical PQR format. (*Source:* Reproduced by permission of American Welding Society, ©*Welding Handbook*).

Clause 4 also provides details about the test plates (Figure 14.9) or pipes that are to be welded for the development of the PQR, as well as for welder qualification. The location and orientation of tensile, bend, and Charpy Impact test specimens to be extracted from the test coupons are defined in this section. Two sets of test coupons are required to generate both longitudinal and transverse bend specimens. The coupon thicknesses and joint details depend on the procedure being qualified. It is common for the test coupons to use complete penetration groove welds, but an additional requirement may include fillet welds, or they may be permitted as an alternative to groove welds. Obviously, the welding parameters and conditions that will ultimately be used to develop the weld procedure specification must be used to produce the test coupons. Therefore, a preliminary WPS is first developed and then verified through testing prior to developing a final WPS. Some contracts may also require the development of joint "mock-ups" for testing that include joint details similar to the actual part or fabrication being welded.

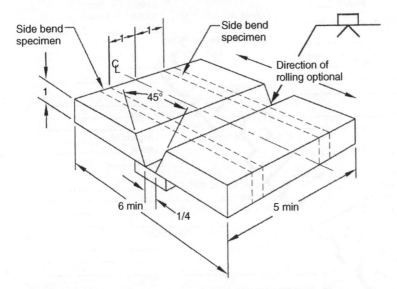

Figure 14.9 Typical test plate. (*Source: Welding Essentials, Second Edition*).

The contract may require other tests such as hardness, fillet weld break, macroetch, elevated temperature, corrosion, and all weld metal tensile tests. Such tests may be needed due to unique service conditions or other factors. Nondestructive testing such as radiography may be required, or in some cases as mentioned above, substituted for mechanical property testing. Acceptance criterion for the various forms of testing is described as well. For example, AWS D1.1 allows up to a 1/8 in. surface discontinuity (such as a crack) on a bend test specimen as long as the sum of all discontinuities does not exceed 3/8 in.

Upon completion of an acceptable PQR, a final WPS (Figure 14.10) can be developed. The WPS is the document that the welder uses to produce the weld. It must include all details needed by the welder such as joint design, base and filler metal type, polarity, shielding gas type, current, voltage, travel speed, and so on. The WPS must reference a PQR that verifies that the WPS procedures and details will produce weld properties that are acceptable for the engineering design of the component or fabrication being welded. The format of the WPS may vary but is considered acceptable as long as it includes the necessary elements the welder needs to produce the desired weld.

WELDING PROCEDURE SPECIFICATION (WPS) Yes ☐
PREQUALIFIED ________ QUALIFIED BY TESTING ________
or PROCEDURE QUALIFICATION RECORDS (PQR) Yes ☐

Identification # ______________________
Revision ______ Date __________ By __________
Authorized by ______________ Date __________
Type—Manual ☐ Semi-Automatic ☐
Machine ☐ Automatic ☐

Company Name ______________________
Welding Process(es) ______________________
Supporting PQR No.(s) ______________________

JOINT DESIGN USED
Type:
Single ☐ Double Weld ☐
Backing: Yes ☐ No ☐
Backing Material:
Root Opening ______ Root Face Dimension ______
Groove Angle: ______ Radius (J–U) ______
Back Gouging: Yes ☐ No ☐ Method ______

POSITION
Position of Groove: __________ Fillet: __________
Vertical Progression: Up ☐ Down ☐

ELECTRICAL CHARACTERISTICS
Transfer Mode (GMAW) Short-Circuiting ☐
Globular ☐ Spray ☐
Current: AC ☐ DCEP ☐ DCEN ☐ Pulsed ☐
Other ______________________
Tungsten Electrode (GTAW)
Size: __________
Type: __________

BASE METALS
Material Spec. ______________________
Type or Grade ______________________
Thickness: Groove __________ Fillet __________
Diameter (Pipe) ______________________

TECHNIQUE
Stringer or Weave Bead: ______________
Multi-pass or Single Pass (per side) ______________
Number of Electrodes ______________
Electrode Spacing Longitudinal ______
Lateral ______
Angle ______

FILLER METALS
AWS Specification ______________________
AWS Classification ______________________

SHIELDING
Flux __________ Gas __________
Composition __________
Electrode-Flux (Class) ______ Flow Rate __________
Gas Cup Size __________

Contact Tube to Work Distance ______________
Peening ______________
Interpass Cleaning: ______________

PREHEAT
Preheat Temp., Min ______________
Interpass Temp., Min ______ Max ______

POSTWELD HEAT TREATMENT
Temp. ______________
Time ______________

SAMPLE

WELDING PROCEDURE

Pass or Weld Layer(s)	Process	Filler Metals		Current		Volts	Travel Speed	Joint Details
		Class	Diam.	Type & Polarity	Amps or Wire Feed Speed			

Form E-1 (Front)

Figure 14.10 Typical format of a WPS. (*Source:* Reproduced by permission of American Welding Society, ©*Welding Handbook*).

WELDER, WELDING OPERATOR, OR TACK WELDER QUALIFICATION TEST RECORD

Type of Welder _____

Name _____ Identification No. _____

Welding Procedure Specification No. _____ Rev _____ Date _____

	Record Actual Values Used in Qualification	Qualification Range
Variables		
Process/Type [Table 4.10, Item (1)]		
Electrode (single or multiple) [Table 4.10, Item (8)]		
Current/Polarity		
Position [Table 4.10, Item (4)]		
Weld Progression [Table 4.10, Item (6)]		
Backing (YES or NO) [Table 4.10, Item (7)]		
Material/Spec.	to	
Base Metal		
Thickness: (Plate)		
Groove		
Fillet		
Thickness: (Pipe/tube)		
Groove		
Fillet		
Diameter: (Pipe)		
Groove		
Fillet		
Filler Metal [Table 4.10, Item (3)]		
Spec. No.		
Class		
F-No. [Table 4.10, Item (2)]		
Gas/Flux Type [Table 4.10, Item (3)]		
Other		

V̶I̶S̶U̶A̶L̶ ̶I̶N̶S̶P̶E̶C̶T̶I̶ON (4.8.1)

Ac̶c̶e̶p̶t̶a̶b̶l̶e̶ ̶Y̶E̶S̶ or NO _____

Guided B̶e̶n̶d̶ ̶T̶est Results (4.30.5)

Type		Type	Result

Fillet T̶est Results (4.30.2.3 and 4.30.4.1)

Appearance _____ Fillet Size _____

Fracture Test Root Penetration _____ Macroetch _____

(Describe the location, nature, and size of any crack or tearing of the specimen.)

Inspected by _____ Test Number _____

Organization _____ Date _____

RADIOGRAPHIC TEST RESULTS (4.30.3.1)					
Film Identification Number	Results	Remarks	Film Identification Number	Results	Remarks

Interpreted by _____ Test Number _____

Organization _____ Date _____

We, the undersigned, certify that the statements in this record are correct and that the test welds were prepared, welded, and tested in accordance with the requirements of section 4 of AWS D1.1, (_____) Structural Welding Code—Steel.
(year)

Manufacturer or Contractor _____ Authorized By _____

Form E-4 Date _____

Figure 14.11 Example of a welder performance qualification form. (*Source:* Reproduced by permission of American Welding Society, ©*Welding Handbook*).

Once the WPS is developed, the parameters and other details describing the welding procedure must be adhered to. However, it is very common for a condition to arise during fabrication which may require a change to the details of the WPS. Some examples include problems obtaining a certain filler material or gas composition, the desire to weld faster to meet production demands, or a change in the base metal type from a supplier. When such changes are requested, requalification of the WPS will be required if the variable that is changed is an essential variable. A variable is considered essential when it is anticipated that a change to this variable will affect the mechanical properties. Otherwise, it is considered a nonessential variable that may be changed without the requirement to qualify a new WPS. AWS D1.1 and other codes provide long lists of essential and nonessential variables. In many cases, changing variables by relatively small amounts is permitted (and therefore, considered a nonessential change) without the need to requalify. For example, in D1.1, for a given wire diameter, it states that an increase or decrease in GMAW amperage of up to 10% is considered nonessential. If the change exceeds 10%, the procedure must be requalified.

In addition to establishing a qualified WPS, it is obviously equally important to establish qualified welders. This is known as Welder Performance Qualification (WPQ) and is verified by a WPQ record (Figure 14.11) that is similar to a PQR in that it includes mechanical test results of test coupons produced by the welder. Welding position is a very important aspect of welder qualification because some welding positions such as flat require less skill than others, such as vertical or overhead. So, if a welder qualifies with the flat position, that is the only position they are permitted to use. However, in many codes including AWS D1.1, welders may be automatically qualified for easier positions if they choose to qualify using a more difficult position. For example, a welder qualifying to the 3G (vertical) position with plate automatically qualifies for the flat and horizontal positions. D1.1 also provides provisions for qualifying welding operators, which refers to welding personnel who operate a mechanized welding machine or robot.

14.2.2 Fabrication and Inspection

Clause 5 of AWS D1.1 is titled "Fabrication" and includes details such as heat treatments, repairs, welding environment, minimum fillet weld sizes and acceptable weld profiles, base metal preparation, the use of backing bars, and repairs. Clause 6 is a very important section for the Certified Welding Inspector and other NDT personnel in that it focuses on inspection. The visual acceptance criteria table defines which flaws or discontinuities are acceptable, and which are to be rejected. This section also provides guidance, requirements, and acceptance criteria for Liquid Penetrant, Magnetic Particle, Radiographic, and Ultrasonic Testing.

14.3 Test Your Knowledge

I) Fundamental Concepts—True/False

The following true/false questions pertain to some of the most important fundamental concepts in this chapter:

1) A code is a form of a standard that includes mandatory requirements.
2) A primary difference between codes and specifications is specifications usually apply to materials or products, while codes apply to processes.
3) The PQR is the document that verifies welder qualification.
4) Once a welder is qualified, they are permitted to use any position, even if they only qualify in the flat position.

5) The WPS is the document that provides all the necessary welding variables for the welder to follow.
6) If an essential variable is changed, the welding procedure will have to be requalified.

II) Solve a Welding Engineering Problem

The following challenge represents a typical problem Welding Engineers might encounter in their career:

You have just been informed that your company is bidding on a new project involving a significant amount of welding on steel pressure piping. Your boss is planning to adhere to the AWS D1.1 code for this project and wants you to confirm that this is the appropriate code to use. What is your advice to your boss?

Recommended Reading for Further Information

AWS Welding Handbook, Ninth Edition, Volume 1—"Welding Science and Technology," American Welding Society, 2001.
Structural Welding Code—Steel, American Welding Society, 2010.

15

Safe Practices in Welding

This chapter is intended only as a brief overview of selected important topics associated with welding safety, and is not intended to be all-encompassing, or to be used for the development of safe practices and procedures. For comprehensive details and training guidance regarding safe practices during welding, the common standard in the United States is ANSI Z49.1, "Safety in Welding, Cutting, and Allied Processes." This standard is listed in most AWS codes and is typically invoked when a welding code is mandated by the contract or law.

15.1 Electrical Shock

Most arc welding power supplies operate at open circuit voltages of 60–80 V. Although these voltages are relatively safe, the risk of serious injury or death exists if proper electrical safety practices are not followed. Common electrical safety practices include avoiding working in wet or damp conditions, use of rubber soles, proper maintenance and grounding of equipment, and special precautions when two or more welders are working on the same structure. Some equipment (such as Electron Beam Welding) operate at much higher voltages than arc welding equipment, increasing the danger if proper electrical safety is not practiced.

15.2 Radiation

Welding arcs produce both ultraviolet and infrared radiation. Damage to the eyes can occur if the arc is viewed without the use of proper lenses. Welders should also be careful to protect exposed skin to prevent painful burns similar to sunburn. Electron Beams and Lasers produce radiation as well. Some Lasers produce dangerous wavelengths of light and require special eye protection. All equipment producing radiation should use appropriate screens or booths to provide additional protection to personnel in the vicinity of the welding operation. Thoriated tungsten electrodes used for Gas Tungsten Arc Welding are radioactive, and therefore, proper ventilation and protection is required during grinding.

15.3 Burns

Obviously, the risk of getting burned always exists when working around any welding operation. In addition to radiation burns, a welder or welding operator may be easily burned by

Welding Engineering: An Introduction, Second Edition. David H. Phillips.
© 2023 John Wiley & Sons, Inc. Published 2023 by John Wiley & Sons, Inc.
Companion Website: www.wiley.com/go/Phillips/WeldingEngineeringIntroduction

touching hot metal or getting hit by spatter or sparks. While arc welding, proper protective equipment including a welding helmet and fireproof protection for the rest of the body must always be worn. Anyone working in and around a welding operation should always assume that a welded part is hot and avoid touching it with bare hands. In addition to fireproof clothing, arc welders should avoid the use of pants with cuffs and open pockets, both of which can easily catch molten spatter.

15.4 Smoke and Fumes

Welding processes produce a wide variety of potentially hazardous fumes. Sources include molten metal vaporization, oils, paint, coatings such as zinc, and fumes and gasses from decomposing fluxes. Proper ventilation must always be used. Other important considerations include the size of the welding space, the type of welding process, the amount of welding being conducted, and the location of the welder's head relative to the flow of the fumes. Both general and local ventilation techniques are typically practiced. General ventilation refers to methods for ventilating the entire space such as opening doors to allow natural ventilation, while local ventilation refers to methods to protect personnel at a workstation such as with a movable hood.

15.5 Welding in Confined Space

Special precautions must be taken when welding in confined spaces such as tanks. Confined spaces produce extremely poor ventilation, and therefore, special proactive approaches to ventilating such as providing oxygen to the welder and monitoring air quality in the space are typically mandated. The Z49.1 document provides specific guidance for safe practices when welding in confined spaces.

15.6 Fire and Explosion Danger

Most welding processes produce significant sources of heat as well as sparks and spatter which can travel up to 35 ft. Therefore, combustible material or fuel should be kept at least 35 ft. away from the welding area. Combustible material on the other side of any wall adjacent to the workstation should be removed as well.

15.7 Compressed Gasses

Pressurized gas cylinders used for many of the arc welding processes must be handled properly to avoid explosions or leaks. Safe handling of compressed gasses is also covered in ANSI Z49.1 and includes labeling, storage, gas withdrawal, valves and pressure relief devices, prevention of fuel gas fires, and concern for air displacement. The concern for air displacement refers to the use of gasses that are capable of displacing oxygen because they are either heavier than air (argon) or lighter than air (helium). For example, when welding with argon, if proper ventilation is not ensured, the argon will begin to pool at the floor and rise in depth much like filling a room with water.

Eventually, asphyxiation is possible if the depth reaches the level of the welder's head. The same danger exists with helium, but in this case, the pooling begins at the ceiling and grows downward. This can be an especially dangerous situation when welding overhead.

15.8 Hazardous Materials

When dealing with hazardous materials, the use of Material Safety Data Sheets (MSDSs) becomes paramount. Typical hazardous materials around a welding operation may include fluorine, zinc, cleaning compounds, chlorinated hydrocarbons, and chromium and nickel in stainless steels. When welding involves hazardous materials, special ventilation techniques are typically required, and the collection of air samples may be mandated. OSHA and other organizations have established allowable limits of airborne contaminants referred to as threshold limit values (TLV) and permissible exposure limits (PEL).

15.9 Test Your Knowledge

I) Fundamental Concepts—True/False

The following true/false questions pertain to some of the most important fundamental concepts in this chapter:

1) Welding arcs produce both ultraviolet and infrared radiation.
2) There are no safety concerns when using argon shielding gas since it is an inert gas.
3) The maximum distance welding spatter can travel is about 5 ft.
4) One source of hazardous fumes when welding is vaporized metal from the weld puddle.
5) The two main types of welding ventilation are known as local and general ventilation.

II) Solve a Welding Engineering Problem

The following challenge represents a typical problem Welding Engineers might encounter in their career:

   Your company is starting a new project that will involve a significant amount of overhead welding of stainless steel using the GTAW process and helium shielding gas. You are aware of a particular safety concern when welding overhead with helium gas shielding. Please explain this concern to your boss.

Recommended Reading for Further Information

ANSI Z49.1, "Safety in Welding, Cutting, and Allied Processes," now available as a free download on the AWS website. 2021.

AWS Welding Handbook, Ninth Edition, Volume 1—"Welding Science and Technology," American Welding Society, 2001.

Index

Welding Engineering: An Introduction, Second Edition. David H. Phillips.
© 2023 John Wiley & Sons, Inc. Published 2023 by John Wiley & Sons, Inc.
Companion Website: www.wiley.com/go/Phillips/WeldingEngineeringIntroduction

Printed and bound by CPI Group (UK) Ltd, Croydon, CR0 4YY

16/04/2025

14658474-0005